Apocalyptic Planet

Apocalyptic Planet

FIELD GUIDE TO THE EVERENDING EARTH

Craig Childs

Pantheon Books · New York

Pantheon Books and colophon are registered trademarks of Random House, Inc.

Portions of chapters in this book were previously published in the following
publications: "Civilizations Fall" in *High Country News* and in *Orion*,
"Sea Rise" in *High Desert Journal*, and "Mountains Move" in *Men's Journal*.

Library of Congress Cataloging-in-Publication Data
Childs, Craig, [date]
Apocalyptic planet : field guide to the everending Earth / Craig Childs.
p. cm.
Includes bibliographical references.
ISBN 978-0-307-37909-2
1. Earth—History—Popular works. I. Title.
QB631.2.C48 2012   550—dc23   2012006012

www.pantheonbooks.com

Jacket design by Linda Huang

Printed in the United States of America
First Edition
2   4   6   8   9   7   5   3   1

For Regan,
the other side of the coin

The word "apocalypse," from the Greek *apokálypsis*, originally referred to the lifting of a veil or a revelation. The common definition as a destructive worldwide event is more recent. In this book, it is both.

Contents

Introduction

I took the idea for this book from my aunt who was sitting on her kitchen floor weeping. We were in her home in Southern California with a broken refrigerator pulled out from the wall, food rotting inside. Streams of sugar ants laced the ceilings and baseboards. It was a seasonal infestation, tiny ants coming out of the ground and filing into her every room. She also had rats. We could hear them padding behind walls and along trusses overhead. I had found one dead in her bathroom vanity. The ants were all over it.

My aunt was in the middle of writing her own book about how to survive the end of the world. Her book was meant to prepare people for what she saw as a coming change, a much-heralded apocalypse and a dawning of a new age. Now she was having trouble just getting through her day.

"I feel awful," she cried. "Everything is falling apart."

She and I used to talk about end-times. That was years back in New Mexico, where she was born and raised almost within earshot of the first atomic explosion. She had looked me in the eye and told me the end was coming and I would be one of those who survived. With a sad and loving smile she had said, "If anyone is going to make it, you are."

Why did my aunt believe this? Maybe because I could skin and eat a snake or knew how to start a fire in the rain. But I didn't know what exactly I was supposed to survive, or even if I wanted to. I was younger when she told me this, a bit more spry anyway. Now

I was too chunky and balding to survive anything of consequence. I couldn't even fix her refrigerator. Nor could I do anything about the rats or ants, nor the wildfires that had begun burning in the hills above her house, nor the tectonic fault itching to slip directly beneath us. Even the divorce she was going through was beyond me. I crouched before her, and all I could do was put my arms around her as tears started again, her life crashing in all at once. Together we weathered the terrible squall, what is said to be the end of the world.

With a strained laugh, she sniffed and palmed away her mussed bangs. "I feel so stupid," she said. I told her it happens to the best of us.

I put her to bed, pillows behind her back, and brought her a bowl of hot soup. When her eyes closed, I took her empty bowl. In the bathroom I doubled a grocery bag and removed that wretched rat, dropping its ant-maddened remains into a trash bin outside. I tidied her sink, quietly closed the door behind me, and made off with her book idea.

The term "end of the world" is thrown around as if we know what it means. Apocalypse? What sort of apocalypse—one that destroys civilization, life, the entire planet? How does it work? Is there a way to stop it, or is it just going to steamroll us?

And are we even asking the right questions?

Most people, when you ask, are a little vague on the details. Informed mainly by blockbuster films, the popular vision is that the end will be rather sudden and accompanied by a thrilling soundtrack as cities slide into the ocean and global climates swing overnight. Sure, that's one way it could happen.

Robert Frost mused it would be by either fire or ice. His conundrum was a product of the nineteenth century. Frost missed the wider, more circular array of options—volcanic cooling of the atmosphere, fossil-fuel-generated warming, global permafrost-methane releases, reentering ejecta from an asteroid impact burning the planet and sending it into a biological tailspin, and so on. Since Frost's time, we've girdled the earth with temperature probes, gravity sensors, and mass-balance buoys. Ice caps have been cored, ancient geographies exhumed. Trace down through ice cores and ancient lake-bottom sediments, pick at fossils and ruined cities, and you will see that the scientific and anthropological records tell a much more complex story.

Like any book, this one does not have only one starting place. As much as it was my aunt, it was also an earthquake. In mid-January 1994, driving back from guiding a trip in Baja, I stopped in Los Angeles for one night. I was sleeping on a couch on the third floor of a concrete apartment building in Pasadena when, a couple hours before morning rush hour, one of the most violent urban quakes ever recorded in California struck the city. Dressed in nothing but a sheet, I was off that couch in about two seconds as the floor banged back and forth beneath me.

This was not a huge event in global terms, 6.7 on the Richter scale, but it was one of the fastest ground accelerations ever recorded on the continent. I didn't have time to rub my eyes or blink. I knew you were supposed to get in a doorjamb to protect yourself, and that is exactly what I did. In the darkened light of the living room, where I'd just been, I could see shelves emptying themselves, books shuffling off the coffee table and onto the floor. This flimsy doorjamb wasn't going to save me if the building came down.

I had never been in an earthquake before, impressed by the sharpness of its pulses as if the building were being rapidly jerked around by its shoulders. This doorjamb seemed no place to be. I darted back into the living room, where I spread my feet and surfed the quake.

In those moments, my picture of the earth was remade. The floor felt as if foot pedals were pumping beneath me, a continental margin humped up on the back of a passing tectonic plate. Humans may have a big hand in carpeting the atmosphere with heat-trapping gases and dumping every toxin we can imagine into waterways, but when the earth decides to roll, it is no longer our game. Right then, this wasn't the planet I thought it was. All bets were off. The ground was moving, and overpasses were pancaking all around the city.

About ten seconds into it, the euphoria wore off. I must have instinctively known the tensile strength of concrete reinforced with tied rebar, and we were near the limit. I remember thinking, "If this goes on for about five more seconds, this building is coming down."

At that moment, the shaking subsided. The tectonic rumble echoed away. I stood slack jawed, amazed. Car alarms must have been going off everywhere, but to my ear there was nothing but unbreakable silence. The world was again still.

With power out, predawn dark was filled with stars.

Fifty-seven people died that morning in the Northridge earthquake. If it had struck during rush hour, that number would have been in the hundreds or thousands, freeways jackknifed under streams of commuters. A couple points higher on the Richter scale and it would have been the big one so many talk about, the catastrophe we seem to dream of as if we just can't help ourselves, drumming our fingers at the edge of apocalypse.

Power came back on between tantalizing aftershocks. I and the two others who lived in the apartment gathered at the television to get a sense of the damage. Seen from a helicopter, it was mostly buckled streets, fallen fronts of buildings, and freeways collapsed in bright morning light. As the helicopter turned to the eastern side of the Santa Monica Mountains, I witnessed something that made the experience viscerally indelible. The mountains on the television suddenly went hazy, as if a single blow had struck dust from a drumhead. I was about to say it must be an aftershock, but I was interrupted by a sound rising in the distance, not from the television, but from outside. The rumble grew as it passed over the city toward us, a wave rolling through the earth's crust. It hit us like a shoulder tackle, and our building violently swayed for a few seconds before settling again. The rumble passed and I understood in my bones the connected curve of the planet. Nothing was separate. We were all in this together, and I mean *all*, everything under the dome of the sky.

That event lifted a veil for me, and I glimpsed a seething and perilous world underneath. I wanted more. What else happens? What massive, elemental changes could come not to one city alone but to the entire planet? This book answers these questions. I will show you what it looks like when it happens. The interest is not merely prurient, the tales not only cautionary. This is what the earth is capable of. It is something that should be known.

To write this book, I traveled to nine different locations around the globe, each an apocalyptic landscape in its own right, an analogue for a likely ending to life even remotely as we know it. If a particular series of events happens, this is what the world will look like. I traveled from the inside bend of South America where sits the driest nonpolar desert in the world to the tectonic madhouse of the Tibetan Plateau to the severe, biotic dearth of central Iowa.

In the original sense of writing a field guide, I kept assiduous notes in journals I carried with me, pages crisped by sun and dotted with rain spatter. This book is what became of those pages. In the most desolate, phenomenal, and downright strange parts of the world, I recorded moments of global regimes shifting entirely. I saw when and where maximums and minimums were reached and new ages began. Land bridges have been consumed by seas, ice sheets have buried the locations of major cities, and formerly serene vistas have been encased in lava and volcanic ash. Comparing what is seen on the ground with geologic records spanning more than 4.5 billion years, I found evidence of an excitable planet where frequent and even regular catastrophes have not taken a hiatus for our benefit. Our version of the real world may be the most fleeting of pleasures.

This much scientists agree on: Five times in the history of the earth, most life has winked out. Five times, one species after the next disappeared, the chain collapsed, grazers died as the plants they depended on were lost, and predators disappeared shortly after, life on earth reaching as close to zero as you'd ever care to get. Up to 90 percent of life in oceans and 75 percent of life on land have been suddenly eliminated. These are the really big ones we talk about, the upheavals that lie at the far end of the pendulum swing, endings you would be glad not to witness.

But you may not have a choice, because we appear to be in one now.

On this much, too, scientists agree: The sixth mass extinction is well under way. Numbers of lost and declining species are rapidly rising with no end in sight. Some researchers offer outside estimates that as many as half of all remaining species may disappear within the next century. Since declines on this scale have happened only a handful of times in the fossil record, this point in earth's history appears pivotal.

Conditions will not remain as they are. That is a guarantee. They never do. In the geologic past, deserts have swallowed the globe, and most of the planet has been infrequently locked in ice, equatorial seas bobbing with slush. In one form or another, these changes have kept going, the earth shifting and jerking between equilibriums, pushed and pulled by all manner of forces, from planetary axis tilting to flushes of CO_2. Our own warm and relatively moist geologic era of the Holocene has gone on for about ten thousand years, and yet it is only a sliver. Current conditions represent about 10 percent of what the earth has been like over the last three million years. Even a mere six thousand years ago, the climate of what is now Phoenix, Arizona, was nearly uninhabitable, water hard to find in the middle of a hyperarid desert.

We fret about whether we will survive the next major change. But who are we talking about being the ones to survive? Humans with our tricky thumbs and extraordinary ability to adapt and spread, familiar ecosystems, or the simple presence of life itself?

What would it mean to be the last ones standing on an ultimately sere and ruined planet?

Just to see it, to taste it, and to experience it firsthand, I stepped forward. Wanting to know what the end is made of, I went there. In the company of lava, ice, and blowing desert sands, I worked to see beneath the surface appearance of things and stand in the presence of apocalypse.

Apocalyptic Planet

1

Deserts Consume

I want to get me a little oblivion, baby.

—Counting Crows

SONORA, MEXICO

A DEAD COW lay on the side of the road, nothing left but hide and bone. We found it along a Mexican two-track where we'd been needing a reason to get out of the truck to stretch our legs. The cow seemed as good a point as any. My wife was driving, and a friend named Devin Vaughan was crammed into the back of the cab. Devin's long spider legs pulled him out as he palmed the door frame and squinted into the turquoise sphere of an arid northwest Mexico sky. The engine ticked in the heat as we wandered three different directions into the desert, each looking for space.

Seven years into a drought, the land looked as if a bomb had gone off. It was the slow bomb of desertification, a withering withdrawal of life, horses the first to go, leaving stout but slender white bones. Cows followed, their carcasses pitched in the sand every several miles, as if they'd gone looking for help and never made it. There hadn't been a lick of rain in more than a year, not enough moisture to split a grass seed. Mesquite trees with sixty-foot-deep taproots were the last of anything to have leaves, dust-green brambles running down an arroyo, hard to tell anymore what was still alive and what simply was. I could see a small ranch in the distance, a *pozo* fenced in around a cement stock tank, gray metal windmill blades creaking in a hot, listless breeze. Nobody was home, hadn't been for a while.

Hide pulled over its rib cage like a stiff tent, the cow had an empty buttonhole for a left eye, and it was still wincing in the sun. I crouched at the carcass, tapped it with my pen, hollow sound, death drum. It had been a bull, not a big-boned pasture bull, but a lithe desert animal that didn't quite make it this time. Devin circled back to me. He wore a loose cotton shirt and had bare, sun-ruddied forearms roped with veins.

"Come on, man, get up," he said to the bull. "You gotta move if you want to get out of here."

When the bull didn't respond, Devin looked up, scanning the horizon. He saw a skim of pear-colored dunes to the west and bony granite mountains surrounded by volcanic blowholes to the east. He liked what he saw. Like me, he was a student of desolation. This bull was a good sign. We had arrived in a year of arid baptism, the heart of a drought, perfect time for two mariners heading for a sand sea. Devin and I had been crossing deserts together for more than a decade. He was an engine mechanic working jet boats for a business he owned on the Colorado River in Utah. When the river season was over, he and I would tumble someplace together, aiming for naked stone or blowing sand, anywhere we could disappear.

"Somewhere close to here, you think?" Devin said.

Looking up at him, I said, "I could leave from right here."

"A few more miles," he said "You name it."

He was thinking of a place to set our main water cache, maybe a black volcanic crater out ahead, half swept with sand.

The plan was to trek through the sand sea of the Gran Desierto, North America's only bona fide erg. The landscape is composed of shifting dunes and sand-swept hardpans extending in a lobate pattern up the inside coast of the Mexican mainland. The sand covers maybe a few thousand square miles. As ergs go, it is a small one (the Grand Erg Oriental, an Algerian dune complex within the Sahara, covers seventy-five thousand square miles). It is hemmed in by the brilliant Sea of Cortez on one side and arrays of black volcanoes and ragged peaks on every other. You can always see your way out, at least if you climb to the highest dune. I'd gone in a couple times before, teased the edges, but never a direct shot to the heart like we were planning. This was no epic crossing to plant our flag on some far shore. Devin and I just wanted to go in as far as possible and stay for as long as we could. Rather than an expedition, this was going to be more of a long picnic—a hot, dry, barren picnic.

Regan was dropping us off. She'd drive back alone, leaving us with twenty gallons of water, along with what we stashed in a few other caches on the way in. She had no desire to join us, liking her creeks and mountain meadows.

I walked a few hundred yards out to some other grizzled thing sticking up from a shallow dune. It was a dog, its pelt stiff and burred with sand. It wore a wry snarl, its teeth exposed. I stared back at the dreadful thing, its water gone, thinking that the desert was unimagining us, unmaking our doings, our sinew and fences. In this drought, only remnants of the last world were left, final pieces being dismantled by the wind. The only living creatures were black beetles the size of matchbooks, last of the scavengers moiling into wind-twisted hollows for scads of blown skin and dry bits of plant. Devin shouted from the truck, his voice far off.

We drove on.

Regan's hands swiveled the steering wheel as she gunned us through miles of dustbins. The truck fishtailed, speed kept up so we wouldn't sink. As soon as we hit hardpan, she pulled a U-turn and stopped.

"Here we are," she said. There was no more road. Maybe we lost it miles back, or it just ran out.

I looked back at Devin. We'd start here.

We unloaded full water jugs borrowed from Devin's jet boat company. After we put them in the scant shade of a leafless paloverde tree, Regan kissed me and asked where we would find each other if things went wrong.

"Someplace not too cold in the winter," I said.

"Southern Arizona," she said.

"There's the border crossing to contend with," I reminded her.

"We might not have to worry about borders," she said.

Regan had been raised by a Korean father who as a boy fled war from country to country in the Far East. He had told me of bombs going off around him as he was hurried onto a boat, of drinking from a stream in which he later found body parts. As an adult he finally located in Colorado to raise a family and live a Louis L'Amour life, just like the books he read to learn English, what it meant to be an American when he immigrated. He stockpiled survival supplies and taught his children how to sustain themselves, how to use a knife, how to shoot a gun. At the age of six, Regan was told that she'd have to be willing to kill to protect her family when things turned terribly sour. Speaking now, she was not alarmed when she thought of such possibilities. She spoke as if out of habit. It was just a question she asked, in case transportation was no longer available and communication went down, you'd have wished you had at least had a brief conversation before parting.

"That spring along the border," she said.

"Quitobaquito?" I asked.

"That's the place."

Another kiss, longer, saying remember me. She started the truck and was gone, her rooster tail trailing her back to the highway, which would get her out of this desert by nightfall.

In galling sunlight, the day was still and silent. Dust settled in the distance, our exit sealed. She wouldn't be back for weeks.

"All right," Devin said, clapping his hands. "Anything's possible now."

We sat out the height of the sun under a naked paloverde tree where red ants motored across many small and fallen leaves. I scribbled in a notebook while Devin sat with arms draped over his knees. He stared into the seam between sky and distant dunes.

"What we need is a good view," Devin said. "Some elevation. Get a lay of this place."

"I think it's just sand out there," I said, gesturing toward the horizon.

"There's a right way to do it," he said.

I nodded toward a metamorphosed range northeast of us. "Spend the night on a summit?"

"Yeah, we can map it out from up there."

We had no maps. That was Devin's call. He liked to go the anti-quated route. No GPS either. Travel was all line of sight, compass if you needed.

Later in the afternoon we carried light packs across pebbled ground to where the horizon lifted into rock. The dry range rose like a sail back, and we stayed out of each other's paths as we climbed, cracking out loose rock, knocking down nicks and shatters. We climbed to an ass-wide saddle and stood on it with shielded eyes looking back at the dunes, from where we stood maybe seven hundred feet off the desert floor.

"We want to be there," he said, pointing to a patch of high dunes maybe forty miles away. They looked like white whales crashing up from a hurricane sea.

I watched and nodded, feeling hot and out of breath. It was about 15 degrees Fahrenheit warmer than I had planned, summer hanging on to fall.

"The star dunes," I said.

"That's what they are?"

A complex of star-shaped dune mountains, they stand out in aerial photographs. "I'd go for that," I said.

We crouched together looking down at this expanse of various shades of tan and brown. If there was a crayon in the box for deserts, this would be the color. You'd use it to fill in the parts of the world that are not ice white, forest green, or ocean blue. Deserts are not aberrations. They are one of the quintessential environments of the earth, a quarter of the planet's solid surface covered by them. They have their own ecosystems and are one of the drivers of global climate patterns, deflecting moisture and sending it elsewhere, kicking up globe-traveling dust storms releasing domes of aridity into the sky. As landscapes, they are tricksters, shape-shifters. Their boundaries are uncertain, uneven. They grow and shrink over time and can suddenly blow up at the edges, leaving thousands, even millions dead, as has happened in the sub-Saharan Sahel. The slightest change in rainfall, an inch or two less in one year, can magnify into failed crops and widespread malnutrition around the inhabited edges of hyperarid regions. For every thousand born in the Sahel, which crosses about ten countries where more than thirty million people subsist on the arid margin of the world's largest desert, at least two hundred do not live past the age of five.

Beneath any landscape is a desert. Given the right conditions, one is always possible. Peel back forest or grassland, exceed precipitation with evaporation, render the ground inhospitable to most life, and voilà, you have a fresh desert. It can happen in as small a space as your backyard or can take up half a continent. Even the largest deserts can be ephemeral. Six thousand years ago the Sahara was not nearly the desert it is today. Monsoons made it home to lakes, marshlands, and dune-sequestering grasses. Pollen trapped in ancient, desiccated lake muds in what is now the hyperarid core of the Eastern Sahara tells of tropical savanna and woodlands formerly existing in that same region. The cause of this change to desert conditions appears to be slight orbital variations of the earth in relation to the sun causing nearby ocean surfaces to heat up. This pushed on atmospheric patterns, flipping the jet stream to the north. Monsoons that had kept the area well watered, supporting extensive pastoral societies at the time, ended abruptly. Within three hundred years, the Sahara changed entirely from marshes, lakes, and grasslands to what we see today, the largest sand desert in the world. Human populations dispersed, many heading to the nearby Nile, where this desert-pushed migration may have been key to the appearance of early agriculture and the eventual rise of pharaonic Egypt.

If heavy monsoons slipped back into North Africa, the Sahara would turn to marshes and grasslands equally fast. Many drought-prone drylands like this can be considered ephemeral landscapes, here one six thousand years, gone the next.

From where I live in Colorado, I have seen red storms billowing in from Utah, boundaries blurring as one state throws itself into another. In twenty-five years of living on the west edge of the Rockies, I had seen few storms like this, and now they have been hitting two or three each spring. Something has changed, a shift in wind and aridity. Rain comes down as red mud, oxidized desert sediment transported hundreds of miles and pouring on you as you drive to town, windshield wipers slashing through blood. Mountains usually white capped in spring have turned pink, a color that absorbs the sun's radiation, which has brought on swift, early-season melts. The snowpack is our largest reservoir of water. By midsummer, when the supply for irrigation ditches below the mountains is usually at its peak, that snowpack has begun dwindling and is mostly gone by July. It used to last until at least August, sometimes September. Ditches now go dry early. Crops suffer, yields decline. This is how a desert moves, creeping its fingers into the land.

The process of arable land going dry is known as desertification. Forty percent of the world's population lives in semiarid regions prone to desertification, places considered arable but could go the other way with little change in conditions. The Gobi and Taklimakan ebb outward, devouring farmlands, dumping storms across eastern China, as much as 300,000 tons of burnt-colored dust descending at once over Beijing, streetlights barely burning through the murk in the middle of the day.* Nine hundred million people live in the most vulnerable of the areas—southern Spain, Greece, Bolivia, Australia, central Asia, the American West. Lines for water are becoming longer, more crowded. Rural women in the increasingly desertified north of India haul water in pails, bottles, and jars balanced on their heads, going an average of six miles a day, four gallons each trip. While groundwater shrinks out from under them and land use strains the soil beyond its fitness, these women just walk farther to water taps, wells, or trucks.

Last update, La Niña conditions for eastern Africa had reached a late alarm stage, drought piled on drought, with a forecast for more depressed rainfall. From the underreported pastoral hinterlands came reports of increasing livestock deaths and migrations of tribes, spark-

---

* Deserts generate most of the world's airborne dust, contributing to a global migration of surface minerals. Dust blowing from the southern Sahara is the single largest producer of iron for the mineral-poor soils of the Amazon in South America. Half of this jungle-fertilizing dust originates in the Bodélé Depression north of Lake Chad, which produces about a hundred storms a year, each sending up a plume weighing as much as 700,000 tons, of which about 40,000 tons make it across the Atlantic to South America.

ing off conflict over dwindling pasture and water resources with host-ing communities. Deeper into the desert, death tolls are unknown. A friend had been traveling in Nigeria, and he came back telling me that one year you're taking pictures of laughing children and the next you go back and most of those children have died.

Crouched on this shadeless ridge crest, I was beginning to feel a little scorched myself, popping Tylenol and gulping water out of a warm metal bottle. I shielded my face from the sun. There was no water but for what we brought. The local water holes, bedrock depressions known as tinajas, had all gone dry.

Devin jumped off the ridge without warning, saying, "Check this out." He climbed down the knife face to a mass of dark feathers jammed in a crack twenty feet below us. Something dead. A big bird. Reaching into it, he pulled out a disarticulated wing pronged with dark feathers.

"What is it?"

"A dead turkey vulture."

"Weird."

"I love this place," Devin shouted, thrusting the wing into the air. "This is where a vulture comes to die."

I ran a hand under my hat, thinking we had already miscalculated on the first day by climbing up here too early, should have read the heat better, waited till dusk. I never did that well in the heat—north-ern blood.

"I'm going to have to stick to night travel when we're carrying weight."

Devin eyed me. Between the two of us, I was the weaker, voted less likely to survive if something went terribly wrong.

Lowering the wing, he said, "I'm not dragging your dead carcass out of here."

A few weeks prior to our arrival, two other travelers—mountaineers, I heard—had gone on a similar trek through here. Within this wave-lashed erg, they became separated. One came straggling out the other side sunbaked and blathering. The other did not. We had spoken with Mexican officials who went out to find the body in the dunes. It was easy to find, they said, a classic desert death. First they recovered his pack. It was discarded and full of water. (He must have realized he was losing strength and had a better chance of getting out without the weight.) Several miles farther, his boots were cast off. (I imagine it was easier to travel barefoot, something you find useful in these soft dunes.) After that, his life unraveled more quickly. They found an empty quart water bottle, then his shirt, and then his pants. His hat was the last to go, his last shade abandoned, maybe pulled off and slipped unnoticed from his grip. I asked how they found his body, faceup or facedown. Faceup, they said, which was somehow a relief to me. He had not fallen over helpless. He had stopped to rest, chosen the place with what was left of his mind. When his final threads unspooled, he was looking to the sky, watching the sun take his last water away. At night, the stars and moon must have been comfort, hurling over him until the sun called him back.

It didn't once cross my or Devin's mind that this could happen to us. We had our own way of travel. They probably had theirs, too, and no doubt started with the same optimism and curiosity, but they were dead and we weren't, and we were going to walk through this place as if it were our own personal sandbox. With enough water to see us through a night, we set a meager camp atop the mountain, mapping our future course, watching the desert below recede into moonless dark. At dawn we returned to our starting cache and there waited for nightfall. We watched the day turn as we sat and napped with our gear. Most of our food had been grabbed off shelves in an Asian market in Phoenix, a hasty selection of tiny dried fish in bags, mashed-up dates, seaweed, and crunchy Play-Doh-colored peas caked in wasabi. We weren't strong planners. We had hard cheese for protein, our supplies basic. Once dark settled in, we loaded up about a hundred pounds each, mostly water, and set off into the stars.

Sand washed on us like waves on a beach: first small rollers, then bigger crests into a rising, open sea. Seams in our packs creaked from water weight, making us sound like wooden sailing ships as we pitched and yawed across the ink-black deep. By the time we were out in farther sand, there was no sense using headlamps, nothing to run into, no hole or rock in the way. It wasn't hot, just warm, the desert letting go of its heat as quickly as it takes it, too fickle to stay one way for long. It was beautiful, the dark stillness, the eager rhythm of our steps. I was as happy as I'd ever been in my life, with a good friend and a long way to walk through abject desolation.

I walked right into the wooden club of a fence post. Letting out a yelp, I half scared myself.

"What the fuck?" Devin said from twenty feet away.

I felt down, fingering through a tear in my pant leg.

"There's a cow fence or something sticking out of the sand," I said, trying to remember when I had my last tetanus shot.

Devin did not stop, mind focused on his own personal night trudge. "Leave it to you," he said.

When I looked up, all I could see of Devin was a shade marching against a star field. I shook the fence post, and it was rooted into hard ground, so we'd come across shallow sand. I flicked on my light and shone it around. Several other ironwood fence posts stuck up at various half-drowned elevations, just a few inches each. I had run smack into the one most exhumed. It had been a corral of some sort, had probably been buried for a while, exposed briefly. There must have been something for a cow to eat around here, somebody's idea for how to make a life. There must have been rain. This was long before the current drought, a previous generation putting up a hopeful sign now covered and uncovered.

I think of the ranch hand who got a job out here pounding fence, riding livestock back in from the dunes, where on good years cows ate dropseed and devil's claw. He must have known that this corral was impermanent, useful for as long as the restless dunes stayed at bay. But if I know him at all, he would have felt his work was here forever, spring rains streaming in from the Gulf of California as far as he could imagine into the future. I think this way sometimes. I believe the lay of my world will remain as I know it today, trees where there are trees, river where there is river.

I panned the light, catching Devin like some strange night creature, his eyes darting away. Grunting under the weight of his water, he waved me away and made a sound that said, *Turn that damn thing off*. I was curious, though, wanting to see more of where we were. I moved the beam in a wider circle, the light dissolving to infinity before riding over a humpbacked dune. I marched to its top and swung the light around, absolutely equal in all directions. I felt as if I were standing on the rippled bottom of the sea.

I flicked out the light. Darkness filled in over the *shoosh* of Devin's boots heading up the next rise, aiming for stars and the trailing footfalls of his breath.

In a Hebrew proverb, God is said to have made the desert so there'd be a place he could walk in peace. I have imagined it like this, the clutter of all of creation with its tangled woods and backlogged prayers cleaned away. Just elemental earth. God strides in sandals, pausing to admire the stars every now and then, slipping his hands into the modest pockets of his tunic as he lets out a holy sigh. We were doing the same, only stumbling blind into sand basins, lifting out each boot as if it were a cinder block. We got up on a rolling deck of hardpan sand and moved more quickly, skating ahead through an immeasurable, sublime peace.

Midnight or later, we moved into deeper sand that rose to a mountain range, an even deeper black lifting ahead of us, winking out one star after the next. We switched on our headlamps and found this craggy spine a hundred feet tall. It was almost buried, sand washed by the wind up the throats of its summits. It was an island in the dunes. Our lights swung across stubborn heads of granite tilted in the gloom until we found a notch to pass through to the other side. We climbed dune slip faces, hauled our water on hands and knees up through the gap, and drifted down the other side. Our headlamps probed farther beyond this range. More sand.

We would use this range as a marker, turning course from here toward the dune interior. The range would be something we could see in the day from miles away. Our first water cache would be here. We dropped on the mountain's pebble-strewn apron. Buckles came off, and we strained our packs to the ground. Feeling suddenly light as paper, I kicked off my boots, threw my socks, and ran around flapping my arms, cackling with delight to be in one of my favorite places on the entire planet. I was tired, though, and that was the last of my energy, so I stopped, panting.

"Stupid monkey," Devin said.

We'd go back for more water in the morning, bringing out enough so we could start setting up a lifeline of caches. A gallon per day per person, no brushing teeth with it or washing salt from our eyes. Only for drinking.

The sand was smooth as skin, its grains small, a texture you could roll around in naked and it would actually feel good, like bathing in silk. For the last three million years it had been blowing inland from the Colorado River delta, an upstream landscape reduced to its smallest, roundest parts, mountain ranges worn down and sent here to be chased around by the wind. Camp amounted to laying ourselves on the ground. I drew a sarong to my shoulders, head on a water bag. There was not even a mumble of good night. Cool settled in, and I pulled up a hooded wool serape. When a wind started, both Devin and I curled against it, tucking in edges around our faces so we did not have to breathe the flying grit.

A few hours later, I opened my eyes. The wind had stopped. I peeled open a space with my finger to see an iridescent dawn. The morning was silent and freshly prepared, our footprints leading here gone. I pulled back my cover of fabric, every move spilling sand as I sat up. Fawn-colored dunes extended indefinitely. They looked like hips and shoulders, navels within navels. You could walk without end in this desert, its every curve and ripple an invitation.

Sunrise was about an hour away. A white hole would burn through the sky. The day would change. For now, nothing moved. The dunes lay still as water.

In 2003, Graeme Ackland, a professor of computer simulation and head of the Institute for Condensed Matter and Complex Systems at the University of Edinburgh, tested out desert stability on a simple model of an earth-like planet facing a sun-like sun. This computerized world was populated solely with flowers, enough biologic feedback to maintain a stable environment. Like kids at an anthill, Ackland and his colleagues poked at their simulation, bumping up solar input—the simple equivalent of global warming—until aberrations began to form on the planet's surface where flowers wouldn't grow. These dead zones they called deserts. Small and roundish, like cigarette burns, these deserts morphed, swelled, and shrank much like our own. And like our own, they collected along midlatitude bands. When heat rose past a threshold, the deserts merged all at once and rapidly consumed the entire planet. Every last flower died. The model terminated.

"I wasn't really expecting anything odd; it's such a simple model with no atmosphere or anything," Ackland told me. He had figured a desert belt would form from global warming caused by increased solar radiation. Instead, he saw newborn deserts that sustained themselves even when the heat was turned back down. Once deserts were in place, they did not simply go away. They kept themselves alive.

In his ensuing paper, Ackland concluded, "At critical points this equilibrium breaks down and the system exhibits catastrophic, large-scale events in response to infinitesimal changes in external forcing." In other words, you pass a threshold and with even very small changes, deserts consume everything.

Ackland said, "It seems that some parts of the world may have two possible natural states, either desert or nondesert. If you could magically cover the desert with plant life, that would be enough to cool and self-sustain it. Conversely, if you stripped all the plant life from an existing 'bistable' [two stable states] region, you would get a desert."

Real-world deserts work in much the same way on a very basic level as Ackland's simulation. If you look down on the earth, or an atlas if that's easier, you will see twin bands of brown desert both north and south of the equator. This happens by differential heating of the earth's surface, warmer around the middle, where sunlight lands straight, cooler at the poles where it comes in at an angle. This sets up an atmospheric circulation known as a Hadley cell. Heat rises from the shine of the equator and the earth's rotation swirls this warm air away toward the cooler poles. As it passes over the tropics, this air dumps most of its precipitation, landing warm and dry around the midlatitudes, where it waves over the sky like a blow-dryer. Here you find deserts: Sonora, Mexico, lined up with North Africa's Sahara above the equator, and below it the Kalahari parallel with the Australian interior.

It is worth noting that these Hadley cells are currently expanding around the globe, blooming poleward. Since 1979, this warm, dry circumference of air has increased by two to five degrees latitude both north and south. These numbers are significant; in the past forty thousand years, Hadley cells are believed to have only fluctuated within a range of about two latitudinal degrees. Now they appear to be jumping. The current shift means potential aridity is moving toward the poles by 140 to 350 miles, like what Ackland saw in his simplified experiment. Basically, if you want to see what your location might look like in only a few decades, check a few degrees closer to the equator.

Dear Arizona: Mexico is coming.

Two days into the sand, I carried a couple quarts of water strapped to my back in the blaze of morning light. It was an exploratory amount of water, enough to last until we got a few miles out and could wander back to our cache by nightfall. Devin and I aren't really fast movers. We like to spend time in places, a day here, a day there. We'd move water ahead later. Now we were taking time to walk through a steady wind, getting a feel for our surroundings. Off to the southwest, I'd see his figure now and then cruising through a dune saddle, his knees gauzed by a swift ground wind. I stopped on a crest, adjusted water straps. From here, I couldn't see a living thing other than Devin.

Some people say they feel small in places like this, their lives seeming insignificant in the face of geographic immensity. That's why I come, of course. But rather than feeling small, I think it's more a total loss of any reference or scale. You are starting from scratch here, seeing the world as it is, not so much as you imagine it.

We left our boots at the last camp, to be picked up on the return route. They proved unnecessary, clumsy in the sand, baking our feet. Barefoot was easier, reading the ground for hardpack, gliding across the surface by touch. The sand burned, but you got used to it after about half an hour, navigating to avoid blistering south faces.

The head of a low dune was curled back by the wind; it looked like a divan, a place to recline for a moment. I slumped into its windbreak, elbow in the sand. On the north side, I dug my feet into the faint coolness just below the surface, a little bit of night still there. A gentle but steady wind strained across the surface. Devin would notice my absence, probably already had, though he couldn't see me. He would stop and wait, taking a break himself. All the years we walked complex, untracked land, we never lost each other. Never had we gotten into any real trouble. We were safe that way, watching for each other over distances you can't shout across.

A single yellow sunflower petal blew over the dune crest. It was captured for a moment in the small, swift drum of air where I was reclining. The petal circled several times, ticking and tapping. So there was life. Somewhere. I wondered, could it be a flower abandoned years ago, buried, dried, and only now exhumed? Or was it grown from a stray rain, didn't get far enough to seed? I almost snatched it with my fingers, but it quickly kept going. I was watching it flutter away, clicking across the sand, when a second petal took its place in the lee. Like the one before, it tumbled several times and then shot off. A third followed, and after that another. I crawled to the crest to look over, wondering what I was to see, perhaps an army of sunflowers blowing in my direction?

Across glinting sand, I caught sight of a train of a few flower parts jogging ahead of one another in the wind. Petals continued arriving in twos or threes or by themselves until every one had been presented. It seemed impossible, something I did not know could happen in the world, a flower anatomically divided but unaccountably kept together. How many miles had it traveled in this fashion?

Next came the detritus of pistils and stamens, an excited trail of nuptials jumping into place. Last to go were bits of dried leaves and stem hurrying to catch up. For a while, I wouldn't move. Why in the world would one sunflower be in the middle of the desert? And how could this have happened to it, caught in a self-organizing wind, an act of chance that seemed highly unlikely. A seed had opened within a vexing drought, then wind carried it piecemeal but intact as if on some magical errand, dried with the yellow still in its petals. Its methodical passage would not show up in a computer model. Statistical improbability, scientists would call it noise, practically meaningless. But it said something about how nature works. I had caught a glimpse of order borne through the chaos of wind. There were angels in its architecture.

Deserts behave in unexpected ways, bouncing weather off their backs, drawing in droughts in one place, sending precipitation to another. What happened to this flower extends to a global scale. A prolonged, river-stripping drought in the desert of Western Australia between 1992 and 2003 resulted in historically unprecedented snowfall in East Antarctica. This hot Australian drought deflected moisture across the bottom of the Indian Ocean, dropping a precipitation anomaly on Antarctica unlike anything that happened there in the last 750 years.

We have lived in calm, pleasant times. Not in modern existence have we seen truly deep or geologically significant droughts. We have been spared, as if a door were being held open for us, a polite gesture. Even the dust bowl of the 1930s that stripped out the high margin of the Midwest was not much of anything, a quick aberration, human suffering rather minor compared with what is possible. What was impressive about the dust bowl, at least in terms of earth's history, was its speed. Due to an unusual combination of factors, fresh deserts sometimes spread across the American High Plains as fast as you can drive. Its speed was an artifact of overuse of the land at a time when agriculture was promoted in every half-arable location, ground moisture sucked from the soil. But humans did not work alone. We never do. A recent NASA study found that between 1931 and 1939 cooler-than-normal ocean temperatures from the tropical Pacific met with warmer tropical Atlantic waters, changing global wind patterns that slowed the movement of precipitation from the Gulf of Mexico inland to the Great Plains. With the Great Depression on, this meteorological signal could not have come at a worse time. It was also unusually warm, temperatures in the Midwest breaking records up to 115 degrees Fahrenheit in eastern Colorado and the Oklahoma Panhandle. Heat pulled any remaining moisture from the ground, and when the wind blew, it took everything. Triple whammy. Which is how catastrophe usually works.

Sediment pelted frames of cars, half burying them in dunes and ripples the color of toast. Windowpanes cracked, and it poured into bedrooms and kitchens of houses where no one lived anymore. Avis Carlson, a popular magazine writer at the time, described what she witnessed as a shovelful of fine sand and dust thrown in the face. She wrote, "People caught in their own yards grope for the doorstep. Cars come to a standstill, for no light in the world can penetrate that swirling murk. . . . We live with the dust, eat it, sleep with it, watch it strip us of possessions and the hope of possessions. It is becoming Real."

Real. With a capital *R*. As if any other pastoral day were not substantial enough in itself.

The fertile American High Plains are sitting on top of a desert. If you fly over dust-bowl country, which is now gridded into green circles of pivot irrigation, you will notice in places expansive, almost oceanic ripple marks beneath a light green cover. These rolling landforms reach out of Kansas and into Colorado, pouring across Texas, Oklahoma, Nebraska, all the way from Mexico into Canada. They are called sand hills. Momentarily pinned down by agriculture or native short-grass steppe, they consist of dunes held in place, some ergs much larger than the one in Mexico. When you drive through them, they seem like ordinary hills, but you notice the way they dip and build on each other, gentle bulges and depressions. Take away the grass, and you see where you really are: in the middle of a sand-dune sea.

Within the last few thousand years, these mid-continent dunes have mobilized at least thirteen times, throwing off their cover of grass during episodes of scouring drought. The biggest single North American erg was not in Mexico, but Nebraska. One Nebraska dune field covered up to twenty thousand square miles, dwarfing the region Devin and I were walking through. It was the largest dune complex in the Western Hemisphere. This did not happen in some distant, geologically incomprehensible past. The last full-dune episode across the High Plains was six hundred years ago. Early-nineteenth-century European explorers encountered remnants, making journal entries on lone dune fronts into which whole rivers disappeared, all of it now humbled back into intensively managed soil conservation districts that hopefully prevent any further dust bowls. Otherwise, Denver may find itself looking across a horizon of wind-drifted sand.

You don't have to wait around to see this kind of process in action. In southern Africa, in the midst of drought conditions and in some places excessive land use, it is happening right now. The surface of large portions of the lower African continent is mobilizing; semiarid grasslands are turning to open sand across Namib, Angola, and Zambia. In the Kalahari Desert, which is more semiarid acacia-riddled savanna than a hyperarid core, climate models anticipate a decrease in rainfall and increase in erosivity, causing grassy dune fields to lose further cover. Unless there is a change to management or weather patterns, over the length of this century the landscape from Angola to Zambia is expected to become a dune zone, a smaller, southern Sahara.

We carried loads of water ahead, going back for more, then moving the weight deeper, creeping our small lives into a glittering field of wind and dunes spooned one into the next. Each cache we dropped along our supply line took us farther in until we were down to about four gallons each, the average that a woman around the Thar Desert in India carries home every day. Only there was no family waiting for us, no little hands to wash. This water was luxury, all you could drink until you ran out.

Did I have to remember what day it was? Even my journal wasn't making much sense, entries blending together, dune easily mistaken for sky. Every evening, a waxing sickle moon grew farther from the sun, bringing a little more light until it felt as if we were walking through a dream, stopping to sleep in a quarter-lit swirl of dunes, Devin up on a crest and me down in the bowl below.

Dawn, I sat pouring cool sand through my fingers. Devin stood in the blue morning dim reading to himself from a book he had brought. It was Charles Bukowski, no sweet love poet. In the first creeping light, thick hardcover opened on his knee—he was carrying a *hardcover!*—Devin recited, "We are here to laugh at the odds and live our lives so well that Death will tremble to take us."

He looked up from his book at the warm dome of light now building in the east. He surveyed the horizon for a few minutes, then said, "Here it comes."

Feet dug into the sand, I took in the last coolness of morning and a last fleeting comfort for my eyes.

He said, "Three . . . two . . . one."

A hydrogen bomb went off before us. Light sheeted across the dunes. Water thief. Blessing, curse. You have to be careful with the sun, warm enough to give life, and warm enough to kill if the balance gets off. The sun's full circle drew into the sky. Within minutes, a warm breeze stirred and pushed past us. It was heat pushing from inland. Ripples of sand motionless all night began moving grain by grain.

Third row back, the earth isn't right up front like Mercury, but we are close enough to be washed in solar winds, orbiting through a veil of superheated plasma. The only reason we are still here is the magnetic field enveloping our planet, a shield against the sun.

"The Oppressor," Devin announced as he stood. He slapped his book closed and turned for the shadowy side of our dune.

Shadows eventually pinched out. We took up some water, leaving most behind for later, and walked into fleets of mirages. Distant dunes rose with the peculiar grace of hot-air balloons, appearing to separate from the earth and drift into the sky. They looked like pale saucers, sun-bright lenses lifting free. Still miles away, these were the heart of this sand sea, a region of star dunes, elevations of sand rising several hundred feet.

Around us, silvery pools began to form. Trickles of what appeared to be liquid mercury flowed into basins, the cool of morning sand inverting on itself beneath suddenly warm air, sun's rays passing through a prism of differentially heated air. It was like walking through a museum of physics constructed by Salvador Dali. Every day like this, every morning physics turned on its ear.

We had pared down clothing; now instead of pants we each wore a sarong, a thin, beach-blanket-sized piece of fabric brought for shade and turned into a better traveling tool. Pants, even shorts, were unneeded, not versatile enough.

We walked up the smooth heads of sand, and we walked down the other side through hard-packed basins littered with thorns and twigs from what had been alive. Around noon, I began shambling, toes dragging through winking and glittering sand. On a sinuous dune crest, I stopped, adjusted a strap across my shoulder. The mirages were dead, chased away by the sun. Midday light was everywhere, not a hint of shade. The sun reflected off every surface. I expected to hear the light sing, a single radiant chord filling the air, but all I could hear was the wind. Sand flustered and twirled. Devin appeared and disappeared in a blustering gauze.

The day's balance was tipping. As it did every day, water rose from me, my vapor becoming traceless. My head was starting to feel like a soft, cooked apple. I could walk another hour at most, but it would be more like stumbling and sliding through steep slopes, a controlled fall down every windward face. Soon, I would have to stop moving. The sky glittered, prismatic with blowing sand, and the ground glinted as thousands of tiny reflective quartz grains rolled and winked their facets. Stars. They looked like stars.

That was enough. It was time for a break. Every day was like this. I spotted Devin, a wrap of fabric blowing around his legs, and I veered left toward him, sliding down off the crest, legs plunging calf-deep as if post-holing through snow.

He shielded his eyes, watching me approach. When I reached him, I unshouldered my gear and pulled out a bottle, shaking it to hear what was left.

"It's hot out here," I said as I sat on my pack. It was a joke, about as funny as I could be.

"And dry," Devin observed.

I drank what was left in that bottle, capped it, and dropped it to the ground. I pulled up the sarong I was wearing, making a tent, a shell against the sun. Devin studied me, saw the redness of my face.

"Why are we out here, again?" I asked.

It was a rhetorical question.

"It looks the same as yesterday," he said, scanning the bulging route ahead. "And the day before that."

"I gotta go under," I said. "I can't be in the sun anymore."

Devin eyed me.

"Yeah," he said. "Let's dig."

This was no tried-and-true desert technique either of us had ever heard of. But it turned out to be the most pleasant option if you're just sitting around in the sun, no shade but for what you can pull over you. We found our places on a clean, steep canvas on the north face of one of the bigger dunes. Where scraggly remains of an ephedra bush grew, we pawed into the sand below it. Streamers and veils flowed down the slope. I crawled into my hole, a sort of grave where I lay with hands working like flippers as I squirmed deeper, covering myself back over, toes and shoulders last, a lizard escaping the sun. We buried ourselves. All that remained was the shade of hat brims, our jawbones half submerged in spindrift, a toe sticking up, the round of a shoulder, as we settled into the small and twiggy shade of this poor ephedra.

Behind sunglasses, I watched the dune edges move. As the wind grew, barrels and curls of sand rotated out of bowls, flaring upward as if from the face of the sun. Within several minutes, any of the cool we might have found beneath the surface wore off as the sand insulated to body temperature. I fell asleep, woke, fell back asleep. Wind sizzled across my cheeks, burying my ears, filling the canals into my skull. Every now and then, I snorfed sand away from my mouth to keep breathing, falling right back to sleep. The caches we dropped behind us were being buried in this wind. But we'd find them, one at the crescent edge of a larger dune foot in line with the mountain range we'd left behind, and another back at the range, marked by a boulder.

I felt as if we were becoming receptacles of drought, our urn-shaped bodies hidden beneath the surface as if preparing for mummification. But we were doing fine, well-watered enough we could last a few days only with what our bodies held. We were water-fat. But even that can leave you. Times do change.

Jonathan Overpeck, one of the world's leading big-picture climate researchers, said, "I think we are significantly underestimating the severity of drought we could get in the future."

Overpeck and his colleagues at the University of Arizona in Tucson have been teasing evidence out of tree rings, mapping a history of extreme droughts in the Southwest. The accuracy of tree rings where squeezed-tight equals drought and spread-apart equals ample precipitation allows a year-by-year reconstruction of moisture conditions for the past two thousand years. What Overpeck and his colleagues found was a clear sign of droughts much larger than anything witnessed in U.S. history. Instead of three or seven years, these lasted up to fifty.

"Let's look at the full range that's occurred over the last couple thousand years and put that on a radar screen," Overpeck said. "Superimpose on that the warming we are now experiencing, because the warming would make all those droughts that occurred in the past pale compared to the future drought. We've already seen in the current drought that's been winking in and out since 1999 how elevated temperatures make a difference. We see a heightened mortality of trees, and we see a faster drawdown of reservoirs."

Overpeck says that if one of these so-called mega-droughts hit in our time, water managers who keep our faucets running would have an unprecedented challenge. "We've never seen anything like these."

I talked to one of those managers, Charlie Ester, head hydrologist for a major water supplier to the city of Phoenix. During a seven-year drought, Ester said he was exhausting all of his climate models. "We don't have anything designed past seven years," he told me. I asked him what impact a much deeper than usual drought would have. "There is no civilization without water," Ester reminded me. "If it does not rain every year and reservoirs go dry and there is nothing to pump, we're moving."

Ester doesn't believe that rains will ever stop completely, but it could rain a lot less than we're used to, a crisis if continued for years or decades. And he was not speaking of just desert cities. Major urban areas in Georgia, such as Atlanta, have recently come to within weeks of draining freshwater drinking supplies, conditions unheard of in the region. Drought reaches far deeper than desert. In 2007, the United Nations brought together 150 experts from forty countries who concluded that if current trends in temperature, aridity, and consumptive land use do not subside or at least level off, within the next ten years desertification will directly cause the uprooting of at least fifty million people who simply cannot remain living where they do. Where will they go? Wherever there's no desert.

In 2011, a year of a debilitating drought in most of New Mexico and parts of Texas, Overpeck gave the U.S. Senate testimony warning of what might lie ahead. In a statement to policy makers, he read, "The current drought, although serious, is modest compared to the magnitude of what could happen in the future." Overpeck told policy makers that even in the absence of climate change, even without human contribution of greenhouse gases into the atmosphere, mega-droughts happen in the normal course of events. So far, we've been lucky.

Of course, this is a rallying cry for climate-change skeptics. The fact that it happens even without us is ammunition for the argument that humans stand apart from climate cycles. But the argument does not hold water. Overpeck says we are making conditions worse, anthropogenic heating making a mega-drought—particularly a really hot mega-drought—more likely.

"We are contributing enough change to the planet that we are moving toward more droughts instead of away from them," Overpeck said. "I'm not running around saying we are going to run out of water. I'm just saying we've got to take the full range of possible climate variability and change into account. Every factor you look at, dust on snow and expanding Hadley cells, there's nothing in there that is going to fight back against droughts."

I had sand in every part of me. My molars wouldn't touch. Buried in it, no longer able to sleep, I watched shadowless hours turn around me as I and the rest of the planet rotated beneath the sun. It was comfortable, hard to stay awake, a little maddening to look across a dreamy, sun-blitzed landscape. Nicks of shadows slowly appeared. They swelled behind each dune. With a caveman groan, Devin pulled himself out of the sand. It was time. He came into my view, dragging out one arm and then the other. Grizzled man wearing a knotted, grease-thinned sarong, he crawled from the earth and roared, lifting himself to his feet as sand poured off him in gallons. Wind swept it up and carried it off.

He stared at his feet, crouched at the hole he just came out of, and shouted through the wind that there was a little scorpion right where his crotch had been. He flicked the thing with his finger.

"Fucker was right on my balls," he said. That was all I needed to hear to get me moving. I had been in long enough. I worked my arms free and after a minute extracted my entire body. Sand poured out of me, kicked upward by the wind, blinding for an instant.

We dug up what little gear we had, threw it over our shoulders, and started back for water. Passing through rounded shadows, I peeled off sunglasses and looked at the world with my own eyes, the landscape leaping out of itself. Just having the sun gone was a sensation you could drink, margarita, on the rocks. The dunes changed color from the muslin blaze of midday. Sun poured into the butterscotch sand as our shadows traveled far. Sunset sand became tangerine, then fell into peach, then salmon. At night, the tipped egg of the moon crossed the middle of the sky. We arrived back where we had dropped our last gallons. In soft ivory halls of light we sat and drank warm water.

In the morning, Devin and I moved one behind the next along high, thin dorsals aiming for the star dunes ahead of us. We were entering the heart of the place. A form of escape? Sure. Here, everything is clean. I don't want to see the weeping mothers of Darfur or the teeming masses of downriver cities worldwide, most people drinking unclean water. There is only so much horror one can take. Yes, I fled. I came to where nothing can be hurt. I moved through one of the most barren chambers of the earth's beating heart, and here I found one of the most erotic places I had ever seen. Dunes flowed behind us. We moved as if through airy fabric, every footstep setting loose teardrops and cascades down sandalwood dune faces. In this desolation, there was no such thing as disaster. You could not kill the wind or the sun or the barren sand.

Hours turned to vapor. I walked in fields of noonday stars where the sky appeared to spread beneath my feet in a seamless, silken sheer.

Full moon that night. The dunes became a milky bath. It was cooler now. I wore socks, missing the warmth of my boots after the sun was down. I prepared my bed in the bottom of a blown-out wind hole, a bright parabola. Careful with the smallest belongings so nothing would be lost, I checked for my pen, journal, headlamp, compass, knife, the only things needed besides water and a little food. Tiny sand crickets were out, their bodies clear as ice, antennae preposterously long. I lay on the wool blanket of my serape and stared up at an oval of stars and the disk of the moon, hung as if in an invisible spiderweb. With my back against it, I could feel the shape of the earth, as crickets mapped my skin.

27

We entered the star dunes like beggars in rags striding into an ivory palace. The sky looked like mother-of-pearl. This far in, we had one gallon left each, two days if we wanted to push it. I voted to stay as long as possible just to soak up the place, the end of our line before having to turn back and pick up the caches left behind us. High-peaked star dunes surrounded us with long, ridgelike arms radiating in all directions. We had seen these from our mountain camp, spotting these white whales breaching in the center of the sea. It was the middle of the afternoon, and I crouched for a while, sarong pulled over my head. It wasn't so hot I needed to go under, or maybe I was getting used to it. But the sun still took it out of me. I hadn't seen Devin for a few hours. We figured we'd find ourselves back at camp by nightfall. I looked up, saw him coming toward me from half a mile away. He looked hot, out of breath. He came closer, taking on size, and dropped down next to me. He had been running, looking for me. I pushed toward him a dead beetle I had found, a little shell of a scarab hollowed out by the wind like a piece of jewelry. It had a shimmering gold color inlaid around the head of its carapace, and I had been carrying it for a couple of hours.

Devin politely admired it and said, "I found something, too."

Unzipping a small leather pouch he always wore at his waist, he dug out a long, curved tooth and handed it over.

A tooth? I turned it in my fingers. It was too big for a coyote or a badger. I would have said wolf, but even red wolves didn't live out here. But it was the only answer I could come up with. It didn't feel quite like tooth, either. The surface was smoother, almost glassy. It seemed half fossilized.

"Where did you find this?"

"There are more," he said. "A lot more."

He gave me his *you'll want to see* look.

I stood. *Let's go.*

About a half mile away I trailed Devin through a gallery of high dunes as he led me into a blowout at the windward base of the largest star dune. Wind had tunneled into the sand-sloped face, leaving a hole at the bottom the size of a baseball field. As we slid down into it, I began finding odd bones half buried, forks of white sticking out. Not heavy like cows' bones, or bulky like horses', they were long and slender like those of deer, only no deer live here. Each was peculiarly weathered, felt almost like porcelain. Whenever I stopped, Devin turned and beckoned me to keep coming. I followed him into the bottom of the hole.

Wind had drilled all the way under the dunes into the hardpan below, revealing a field of bones, an ossified junkyard. You could not take a step without crunching down on one. Moving out ahead of me, Devin turned back and said, "It's like a wave out at sea dipping so far down it exposes a shipwreck on the seafloor."

What was this shipwreck? We were looking back in time. There were bones of all sizes, splinters of rodent legs and ribs beneath the broken blade of a scapula big enough to have come from a sloth or a bison. I picked up dice-sized carnassial teeth, which looked as if they were from cats, big cats. Mountain lion? Cheetah? I lifted them and dropped them, so many bones they spilled from my hands. We were in the remains of an ancient water hole that had dried up. This had been like an African savanna, with a little Great Plains bison thrown in for good measure. I could see the mud in this crackled surface, mud that must have smelled like corpses as the last pools emptied into the sun. Based on the sheer number of bones, it must have been a rapid process. Many animals had been caught in the act, and in a much older and different world.

As we rolled finds between our fingers, a prehistoric bestiary began to take form—predators' cheek teeth, jaws of woolly-fat rodents. I dug out a fist-sized vertebrae, and the high wing of its spinous process suggested camel. I once excavated the bones of an entire Ice Age camel from a wet cave in Colorado, two seasons of digging to get the soggy thing prepped for casting. I had spent days at a time on a single vertebra. The one I dug out for a museum in Colorado had died from falling in a cave, a rather pedestrian accident. The one in the dunes had died with many others, a meteorological event, something beyond daily life and death. Everything was disarticulated, and I saw vultures tugging and feeding among rot until they too had to either leave or die. Ravens might have stuck around to peck at beetles.

This was older than Overpeck's droughts. It looked Pleistocene, dating to the last ice age or even before. In this deeper past there were droughts we could barely comprehend today. A drill core of lake sediments recently taken from within a dormant caldera in northern New Mexico shows droughts that lasted up to ten centuries. A team of paleoclimate researchers led by Peter Fawcett of the University of New Mexico found in this core layers of mud cracks from when the lake completely dried during climatic conditions very similar to our own Holocene, evidence that droughts could right now go beyond even Overpeck's warnings. In a press release following the findings, Fawcett said, "We won't know for sure if it happens again until we get there, but we are certainly increasing the possibility of crossing a critical threshold to severe and lasting drought conditions." By "we," he means our anthropogenic effects, which appear to be setting the stage for conditions that could be far worse than what left these bones in the dunes. Looking ahead, Fawcett and his colleagues wrote, "Models of climate response to anthropogenic warming predict future dust-bowl-like conditions that will last much longer than historical droughts and have a different underlying cause, a poleward expansion of the subtropical dry zones."

About 100,000 years ago, at the end of the last interglacial period similar to the Holocene, droughts like what Fawcett and his colleagues discovered dominated parts of the planet for extended periods, leaving extinctions in their wake. Half of mammals weighing over eleven pounds disappeared worldwide. Maybe that is what we were seeing in this blowout, the remains of one of these ancient extinction-pushing droughts.

The wind had taken the site down about twenty feet through several layers of strata, each layer marking a different event. This hadn't been a single drought, not one last bang and then it was all over. These were horizons of dying fields piled on the next. Survivors had been able to get their numbers back only to arrive again at some other desperate moment when the next shoe dropped. The place had happened in waves until finally no water remained and no animals returned. Sand blew in and covered over the boneyard like a sympathetic hand, a decent burial. I saw the last vultures teetering away in the wind as hides tattered and blew. Sand filled the last carcasses, skated over disarticulation, and finally the delicate curve of a dune rose upon the whole mess. By some accident of wind, the past was exposed, a record of drought upon drought, and in between maybe there were times as fine as our own, long enough you might have forgotten such ruin ever happens.

Devin took to calling the site Olduvai Blowout. "A hell of a lot of dying," he said. For hours we stooped in the sun, no shade but for hat brims. Time was bending like light. One step was worth thousands of years as I wandered among countless animals, the stench of these apocalyptic die-offs long since evaporated.

Devin snatched up something small and let out an "Oh, man, look at this thing."

He tossed it over and I caught it, cradling in my palm a raptor pellet, a half-fossilized glop of bird vomit. It was a perfect little pendant of mouse bones and who knows what else regurgitated by a hawk or an owl long ago. It implied a perch, leafless limb of a snag tree, or talons wrapped patiently on the roost of some beastly rib sticking up from the ground.

Devin said, "You could make a necklace out of that thing."

I imagined it hanging around his neck, something even a grocery clerk would squint at. Devin would say, *I wear the memory of long-gone ages.*

The longer ago something happened, the less resolution we can see. This is the nature of time versus scientific observation. Further back in time, evidence decays, disappears. Puzzle pieces go missing. Methods of study begin to rely on lasers and gas chromatographs rather than a walk in the field, because you simply cannot see the detailed actions of the longer past with your own eyes. You turn to tree rings, which preserve a few thousand years of good records of dry and wet seasons, accuracy sometimes to the exact year or even month. Fawcett's search for droughts 100,000 years ago, beyond the range of tree rings, relies on pollen counts and mud layers at the bottom of a lake and is a little more hazy. Looking back even further, you lose precision as if looking through layers of older and older glass.

Eventually, you can't make out individual droughts, and all you get is the gross history of the earth. Every once in a while, though, you hit a geologic marker, a sign of catastrophic decline. Go back 250 million years, long before dinosaurs, and you run smack into a wall of drought so global and severe that we would be talking not the fall of civilization but the end of it, as extreme as you could get without this planet turning into Mars.

Before this wall, which is known as the end-Permian extinction, the planet was amply populated with trees, ferns, hundred-pound amphibians, lumbering reptilian therapsids, and the usual million or so species in between. Then heat grew. Species declined. Forests were pushed north toward the poles by what must have been runaway Hadley cells, the atmosphere boiled by tropical heat climbing rungs of latitude. Chasing cooler environs, these last trees eventually ran out of minimum daylight, and many pinched out of existence, their genetic lines ended. Most landscapes converted into erosion factories, rivers ran thick with sand and mud, draining off what appears to have been denuded landscapes. Dumping their loads of sediment into warm, evaporating seas, they left fossil records of extreme erosion. Hydrogen sulfide bubbled out of those seas, while exposed salt flats released deadly halogenated gases, as if the earth were heading back to its primordial, Hadean roots. This is the largest die-off the earth has ever seen. Even the oceans became deserts in themselves as seawater turned anoxic, chemical regimes unbearable. To see an analogue for how ocean water turns deadly, look to our own time. Veritably lifeless, low-oxygen regions have appeared and are expanding right now. This is happening most notably along the Gulf Coast of North America, where what is known as a dead zone has grown to more than eight thousand square miles. More than four hundred dead zones have been identified worldwide, some caused by natural, seasonal fluctuations, many from human effluents. Meanwhile, seawater overall has increased in acidity, causing massive die-offs in coral reefs and threatening all manner of water species unable to survive shifting pH balances. As oceans absorb increasing levels of CO_2 from the atmosphere, acid levels rise, dissolving calcium carbonate shells of marine organisms in the water. This is the kind of growing death Ackland saw in his experiment, burn marks expanding across the planet.

Why this happened at the end of the Permian is one of the more pressing questions in earth science. An event like that requires some forcing mechanism, a trigger, but that smoking gun has not been found, only many smoking guns, as if the world were riddled by stray bullets so long ago. The most obvious background culprit was plate tectonics ramming global continents into Pangaea, one warm supercontinent. The earth was experiencing one of its greatest episodes of volcanic activity at the same time, mostly coming from flood basalts in what is now Siberia. A deep layer of char from these volcanoes was deposited in seas immediately before the mass extinction. Meanwhile, CO_2 balances in the fossil record appear to be completely off the charts, carbon liberated into air and sea at levels 10 percent higher than today. In turn, oxygen content reached a low that would have left you straining for breath even at sea level.

What stopped the planet from going over the edge into full extinction is unknown, but over the next ten million years biodiversity slowly returned to previous levels. Continents tore apart from Pangaea and moved into new positions; ocean currents and heat exchanges entered back into fruitful alignments. An incredible excess of atmospheric CO_2, which marked the length of the mass extinction, settled back down as plant life generated oxygen and glacial cycles kicked back in.

Accurate parallels cannot be drawn between the end-Permian and today. There is simply not the resolution from 250 million years ago, and scientists do not have a current data set long enough in time to compare. All we know is that it has happened in the past, Ackland's experiment made real.

We retreated from the star dunes back along previous caches, digging up water as we went. Days later, we found our half-buried clothes and boots. Passing summits half swallowed, we walked out of the erg and into flared creosote bushes, their brittle, wax-crackled leaves turned ocher.

At dawn, we traveled east in Benedictine silence across a pebbled hardpan near where we had seen that first dead bull. We kept going, dunes behind us and ahead lava rock and weeping ironwood trees. Bound in boots, my feet felt awkward, almost suffocated. My joints had widened, toes spread to grip the ground, and the shoes barely fit. In the first clear light of the sun, which had gone from oppressive to welcomed, we began finding seashells and the bones of sea turtles. They lay about us like shaped ceramic tiles, white as snow. About twenty miles from the coast, these were left here by people.

I picked up a smooth *Glycimeris* shell, rubbed its inside with my thumb. It had been glassy once, polished and gleaming from the sea. Now it was pitted and almost chalky. It had been here for centuries.

Devin knelt with one hand on the ground. "Man-sign," he said.

He handed me a curve of broken pottery. Proof. It was brick colored and flecked with shiny mineral temper. The glitter could have been river sediment. From Phoenix, I guessed. Salt River, or maybe the Gila. Prehistoric technology, open-fire wood kiln. It was probably ninth century A.D., the temper having come from hundreds of miles away.

"Shell traders," Devin observed.

I had been studying the pre-Columbian archaeology of this region for several years, had learned some of its old stories. People once carried shells out of this region by the millions. They'd cross the dunes to get to the sea, Adair Bay on the other side a prime shell spot. From the shimmering blue Sea of Cortez, they carried their shells inland, trading them as far away as Oklahoma.

Devin picked up another piece of pottery, a lip from a beige jar marked with lines of faded yellow paint. Wagging it in the air and looking around, he said, "Of course this is where they would have set a camp. They would have kept water out here. Ceramic ollas."

Devin dropped his pack and began pacing the sandy edges, looking for more than just random scatter. He wanted to find where they slept and ate. I did the same, and we fanned out, thumbing through shells and some chipped stone. I crouched at every piece of pottery, eyeing each edge. They were all very old. The people were likely Hia-Ced O'odham, desert dwellers, known as Sand Papago, sand people. Some of their pottery was thin and bore the most subtle of curves. Holding out one smooth piece the size of my palm, I could see the vessel's original size. It had been a water olla, a globe-like jar big enough that I could have wrapped my arms around it. In it they kept the one thing you need to live out here.

The tinajas they would have used were empty for the moment. The largest of them, Tinaja de los Pápagos, "Watering Hole of the People," usually would have at least a little tepid green-black water come fall. Not now. It held nothing but crisp cockleburs and flecks of algae, smooth bedrock hot to the touch. This would not have been a good time for the Sand Papago, a nomadic people who would have probably evacuated the region for the north, retreating to ranges in what is now the United States, locations with a little more water. But those places would have become busy, and right around the time of a particularly difficult drought in the thirteenth century the entire Southwest went through a societal collapse, infant mortality skyrocketing, violent conflict at its peak, villages left empty, bones on the ground.

Drought has never been easy. Everyone eventually feels it.

"Over here," Devin called.

He lifted from a sandbank a round piece of polished basalt the size of a bathroom sink. It was the shape of a sink too, hollowed down in the center to a hole. He hoisted it to his waist to show me.

"Done found a gyratory crusher," I said, surprised. I had only seen them in museums and old pictures. "Those things are rare."

"And awesome," he said. "Look at this." He jerked it over his head and peered up through its hole into the sky, admiring the precise work. This was no rough toolmaking. It was a piece of art. The hole was where milled foodstuff (probably grass seed and native desert legumes) would have fallen through, pulverized by a basalt pestle we could probably find if we looked hard enough, if it were not covered by sand. People had probably been working grass seeds between February and April. In the winters, they would have pounded the hard, slate-colored beans of paloverdes and the smoky seeds of mesquites. Their hands were here, phantoms left in artifacts. This had been important, a precious object nomads could not carry with them. It meant they returned to this place again and again, families and generations, children padding their bare feet in the dust. They probably saw drought many times, and lived.

China has built walls of trees to hold back dust storms spreading out of the Gobi and the Taklimakan. Meanwhile, the Kalahari Desert of southern Africa remains undecided, grass seeds still viable, dormant for the next rain, when they may throw their weight into stopping an even deeper desert from forming. In 2009, the African Sahel, struggling through decades of devastating aridity, was inundated by monsoonal rains. Though the following year the region was hit by the now infamous drought of 2010, triggering large-scale famine, the heavy monsoons suggest the future is still wavering, not yet settled into a continuation of the Sahara Desert. The hyperarid outline of the Sahara became smaller between 1984 and 1997 even during jaw-dropping droughts. Between 1982 and 2002, satellite images have shown the edges of the Sahara actually regreening. As I said, deserts are tricksters, not to be trusted.

We are not on a one-way trip to a brown and sandblasted planet. The past does not tell only one story of our future. There are certainly trends, but along those trends you'll find a roller coaster of ups and downs, indicators hitting highs and lows. There are tipping points and turns, within them aberrations and back pedals where you could suddenly get ahead, throwing grass seeds, spreading rainwater. The point of no return is a moving target.

Around the edges of the Thar Desert of India, grass seed is being dispersed by government mandate to hold the ground, capping dust storms much the way grasses keep the American High Plains from blowing away. Curiously, this grass seed has proven to grow better when scattered by hand rather than by aircraft. The potential scale of human involvement is not always driven by the largest and most efficient technology. Survival and continuity in the desert often require small actions. In that increasingly desertified part of western India, there is a tradition among pastoral villages of preserving sacred groves known as *orans*. Often protected in the name of a local deity, these *orans*, often at least several acres each, are populated by remnant species of shrubs, grass, bamboo, even banyan, said to protect both living and nonliving beings. In places, they are the last pieces of a former land, genetic banks, arks of living species ready to spread between the ups and the downs of drought.

Closer to home, I have friends living on the outskirts of Tucson who have coaxed their water table fifteen feet higher with their own hands and shovels. Building up subtle contour traps, they massaged the ground to the point where you would hardly notice, just another piece of desert property, but where rainfall moves differently. It sinks in places, catches underground. While aquifers have plummeted in the surrounding, developed desert—acres of land actually cracking and slumping into fissures from industrial water removal without recharge—these people have been able to increase their underground storage, and their property bears the appearance of a small desert refuge. This is called survival.

Our last night, Devin and I camped at an eruption crater, a mile-wide hole sunk into the ground where ragged cliffs descended eight hundred feet. As night came on, stars beginning to resolve with a late moon far past full, the crater below us became one emptiness inside another, dark within dark. It is one of several oversized craters in the region, likely caused when magma rose up and hit groundwater about four thousand years ago. It would have been a formidable and sudden event, a cork popping. It was the kind of eruption that could have shot rocks into orbit. A meteorite you see at night as a shooting star could be from this hole. I have wondered if early people witnessed the eruption. The archaeological record says they were around. What a spectacle it must have been, like a nuclear blast, a flare on the horizon and out popped the instantaneous insides of the earth.

Our fists rummaged through the bag of tiny dried fish with crisp and bulging eyes, digging up salt from corners. Finally we passed out as if drunk, food bag left open on the ground, who cared if it blew away, we were gone tomorrow, our ride coming back for us.

In the middle of the night, I sat up and saw a shocking blood-red sky in the north. I had to stare for a moment, trying to shake off the last of a dream. Stars were nearly blotted out by the light. Was it some distant fire? No, not enough to burn out there. For a quizzical half second, I thought it was a huge, brimming volcano. Maybe I was one of those to witness something unheard of, and I only had seconds to live before it hit.

It could have been a nuclear blast, I supposed, the United States finally cutting its losses at the border.

"Hey, wake up."

"Unh," Devin said, then, "Holy shit."

We had both seen odd illuminations in the north. The border is its own special war zone. On the other side are U.S. bombing ranges and military maneuver grounds occasionally lit up with rockets or flares, but nothing like this. After a few minutes, we could see the light wavering, shifting slowly against remaining stars, gyrating over the earth. It was definitely not from an explosion, not one on this planet at least. It was set higher in the atmosphere, something that must have been several hundred miles across. It was a natural event, much larger than anything humans could pull off.

"Aurora borealis," Devin said. "It couldn't be anything else."

Northern lights seen from an unusually far-south vantage.

Later we would learn this was true. A coronal mass ejection from the sun reached the planet at that same hour, a burst of solar wind barraging northern skies with subatomic particles. A time or two every year an electromagnetic back draft hits the upper atmosphere especially hard, and the lights have been reported in Mexico. Usually, it's just a glimmer, the faintest red hat on the horizon. This was a dazzling light show, blood-colored fingers passing across each other. Beneath it, I felt naked, exposed. Our silhouettes were witness to some final glory, a pair of travelers with grandstand seats at the end of the world.

2

Ice Collapses

*After all, time has no physical basis. We can't feel or touch it.
Yet there's almost a sense that we can see it.*

—Chris Turney

NORTHERN PATAGONIA, CHILE

IN THIS EARTH lies a balance, a weight ticking back and forth. You can see it in the ice, glaciers moving out and back like waves and tides; hours, days, months, years. When the world changes, so does the ice. When it changes fast, we call it catastrophe.

There is something haunting about a place where the earth is laid bare, where catastrophe has struck. It is so desolate and exposed it feels as if anything could happen, and as if everything already has.

In the back bowls of the southern Andes, bone-white glaciers drape across mountains. They flow down in crumbling fingers from the great white mass of the Northern Patagonian Ice Field. Mountains are ripped open, summits cleaved to pieces, hulks half gone and pulverized. Bedrock not exposed for thousands of years is being revealed in these decades of subtle, erratic, but measurable warming.

My every footstep kicked up puffs of gray dust that came from the sheer tonnage and motion of a receding glacier, the ground milled into fine powder mixed with broken boulders and pieces of granite. The glacier had been here only a few years ago, its weight grinding over this spot where now there has not been enough time for grass to grow or wind to winnow the fine dust.

The sky was clear, a searing blue unusual for this often cloud-covered region of northern Patagonia. In a cool wind I walked across a sun-warmed surface, every now and then hearing an explosion in the distance, seracs of towering ice falling into this drumlike valley.

The largest seracs coming down were the size of buildings, fronts of glaciers peeling away and exploding into bald, sculpted bedrock couloirs. This was one of the warmest seasons recorded in Chile, and the ice was coming apart.

The glacier I was walking through, the one no longer here, had existed recently enough that the surface was still dewatering. Gin-clear streams emerged from the ground and ran across bare rock.

Another boom rang out, another serac fallen. Usually, you think of global ice loss as purely an issue of melting, but the principal way ice mass disappears is through sheer physical collapse, glaciers and ice shelves fracturing deep inside and breaking apart. Titanic forces were colliding around me, the last of the Ice Age in retreat, warmth settling more deeply year by year, speed of the change picking up to the point it is visible to the human eye. The earth's axis, in its many-thousand-year wobble, has been edging temperatures up since about eighteen thousand years ago. Most of the ice that used to exist is gone, and midlatitude landmasses, such as the bottom half of South America (as well as the Himalayas, Alps, Rocky Mountains, Cascades, Mount Kilimanjaro), are losing ice faster than anywhere else in the world. Of 150 glaciers originally named in Montana's Glacier National Park in 1850, only 26 remain, as if ice loss were sprinting for a finish. This is not happening at the incredible volume of polar regions, but what these midlatitude locations have is going much faster and not being replaced.

Looking ahead, I could see the bold white cap of the Northern Patagonian Ice Field. Along with its slightly larger twin, the Southern Ice Field, this is the largest nonpolar ice mass in the world. The two squat down on the southern Andes, forming eight thousand square miles of white. In places the ice is six thousand feet thick. It broke like a wave through the highest peaks, its rumple-backed body pouring down into glaciers that fingered into their own valleys below, glaciers all making a spectacular retreat.

About a quarter mile from the face of the nearest fallen glacier, I moved down an elephant-gray gully and came upon a piece of lop-sided white shrapnel. It was the size of a two-door sedan, and it had rolled and drifted through a landscape changing sometimes faster than you can turn your head. This was where it ended up, probably its last stop. Concentric circles of drip marks in the wet grit suggested this chunk had more heft when it landed here. I walked the outermost drip marks, pacing out what must have been as big as a bedroom maybe two weeks ago. Placing my hands against its melted, vermicular surface, I was half surprised to find it cold.

Was this some poor, dying wastrel, or was it getting what the ice always wanted, turned back into liquid after thousands of years of being chained into a molecular solid, now freed drip by drip?

Kneeling on damp ground, I glanced back to see if anyone had followed me. It seemed as if I had slipped the rest of the crew. We were making a documentary film, and soon enough they would find me and have me pose by this ice, setting up a boom and multiple camera angles while debating whether I should wear my hat, or how to handle the label on my pack and jacket; was it a sponsor or not?

Our film was one of those save-the-earth flicks, the kind you would see at a Mountainfilm festival. Never having been involved in such a project, followed by cameras through the wilderness, I found myself disappearing from view as often as possible. Last I saw them, they had kicked open their tripods on a barren hillock and were shouting from a hundred yards away for me to put on my blue coat so the lens could find me, my usual drab clothing blending in too well with the background. That's when I dropped into this gully and found this piece of ice that I now had both hands pressed against. I was here for the ice anyway. Saving the world? You can always hope. But to be alive in the last geologic moments of ice, wouldn't you come and put your hands against it? As it tinkled and cracked in the sun, I snapped off a tab and crunched it in my mouth. It turned to water instantly, as if it had been waiting a hundred centuries for this moment.

At best, ours was a half-bungled seat-of-the-pants venture. It started coincidentally with a massive earthquake that ruptured supply lines into southern Chile. Registering 8.8 on the Richter scale, this was the sixth-largest earthquake in recorded history. Last of the crew to arrive, I showed up on one of the final flights into northern Patagonia, leaving Santiago just hours before the quake struck. After that, no more planes. Patagonia was cut off, and a fair portion of our camera gear didn't arrive. We went ahead with whatever we had, but from there on out we were a traveling junk show. Gas was hard to buy. Food lines were crowded. Once we finally got on the road, we ended up leaving an equipment trailer stranded a hundred miles down a dirt road after a wheel broke off its axle. Past that, we lost some of the crew who took off with half our food and gear, giving up on the project entirely because, as the crazy Chilean woman shouted at us, we were acting like cowboys.

We were traveling by boat and foot into the ice-headed watershed of the Rio Baker, one of the largest rivers in Chile, where we would switch from crampons to boats and run the river down to the sea. The point of the film was to protest a series of hydroelectric dams planned for this remote region, an arrival of industry that would dramatically change one of the last wild, underpopulated temperate regions in the world. Valley by valley we crossed the front of the ice field and its many glaciers, dipping at times into old forest. A horse train followed a few miles behind us loaded with an impressive assortment of gear, video equipment, and enough rum to kill a busload of schoolchildren. We couldn't fail.

We moved through heavy woods on what you could barely call a trail, and I looked up for ice, thumbs hooked through shoulder straps on my pack. I searched holes in the canopy as we crossed boulders washed by cold, clear streams of pure melt. We danced and did not pause, boot tips on rock tops followed by a quick teeter across felled wet timber and a long jump to a spongy bank. It was practically a jog. We had a long way to go before the sea, every day having to stop for the shot list, the quotation, the stupid antics of whoever got in front of a camera.

It was early March and warm, end of the austral summer, which meant the melt was on from every ice mass south of the equator. If you wanted to see it go, this was the time and the place. Moving up the Soler Valley, an east-flowing drainage off the ice field, we carried packs in a steep and forested landscape. The forest was dominated by the *Nothofagus* genus, towering deciduous coihues (pronounced "coy-ways"), what you might call beech trees. Through holes in the coihue canopy I glimpsed horned mountains and their white robes, making me eager for this expedition to get to more ice.

We moved through half-lights and shades, following a local guide, Jonathan Leidich, a brawny, ridiculously handsome Colorado-born expat who had come late to our crew. After we lost those first few, who marched away in frustration, we ran into Jonathan at his garage in the small town of Puerto Bertrand. He took pity on us, believing in our cause of protecting this landscape. Without him, we'd still be slogging in marshes and braided riverbed twenty miles back. He knew the way, a footpath through passes and valleys, an old stock route about as wide as a goat trail. He had been out here clearing it himself with machetes, chainsaws, and shovels, carving a path into the wilderness.

The forest drummed with the sound of a woodpecker. Jonathan stopped. We slowed behind him. He was hardly winded, while I had to hold in my breath to listen to what he was hearing.

"There it is," he whispered.

A bird with a blood-red flame on its head came into view around the column of a coihue trunk. Its head hammered at every turn as it scratched toeholds, picking out beetle grubs. Jonathan gestured for us to be quiet as cameras were unsaddled. He whispered its name over his shoulder, *Campephilus magellanicus*, Magellanic woodpecker.

"It is a southern relative of the extinct ivory-billed woodpecker," he said under his breath. "God's bird."

Its thrumming wings carried it down the steep valley and out of sight. Jonathan waited, listened, heard nothing. He seemed satisfied and without another word returned to his swift pace along the trail.

Had we been here twenty thousand years ago, it wouldn't be like this at all. There were no birds here, no trees. All you might have heard were pings and cracks of ice as you were buried under at least a mile of it. I couldn't help feeling the weight, seeing the steely dark interior, our bodies encased like cave people awaiting thaw. It was the last glacial maximum, a time when this part of South America was completely capped from Pacific to Atlantic. Only the occasional high summit would have stuck through, the dark tip of a nunatak piercing a plain of seemingly infinite white. Most of North America was covered at the time, as well as northern Europe. The British Isles were nowhere to be seen. It was a colder time in earth's history, and with enough moisture to lay ice on thick around the world. Now we have come to the opposite end of that swing.

Instead of an unfathomable tonnage of a glacier, I moved through hangdog flowers in crowded underbrush. Old trees blocked the sun. The trail switched down to the valley floor, where we broke into clear, unseasonable sunlight, emerging onto a naked, mile-wide gravel bar braided by a glacial river. For the first time all day, we could move freely under open sky, no more trail confining us. Like bees let out of a hive, we scattered. At the high end of the valley, the ice field itself rose a breathtaking white as it broke across the mountains.

"Awesome!" someone shouted into the crystalline sunlight.

Someone else let out a whoop.

Warmth glimmered off bare gravel. It smelled like rock and sunlight, clean as a desert.

Sledge-shaped mountains with snow-streaked summits stood high in the back of the valley, impinging on the sky. The day was mid-eighties in the sun, and seracs were going off like fireworks. White flares appeared miles away as glaciers crept forward, their edges breaking off and spilling into steep bedrock couloirs below. They looked like comet tails coming down the mountain, disappearing far from our view.

Timmy O'Neill, a rock-star climber, danced up behind me. In a comedic narrator's voice, he said, "Craig wonders how long will it all last? What does it all mean?"

I hadn't known Timmy for long, met him for the first time on this trip. I took him with a grain of salt. He knew what was in my head because it was already in his. He was the expedition jester.

He was right about what I was thinking. You couldn't help it under these curtains of glaciers, with a highly eroded valley floor choked with more glacial debris than the river could possibly move on its own, mismatched boulders tumbled all over one another. This was not old debris. Everything was fresh. Something cataclysmic had happened, signs of flooding everywhere, and not an ordinary flood but a mass of water at least twelve feet deep and a mile wide. It must have looked like a rushing lake when it came through, a part-liquid and part-solid surface rafted with uprooted trees.

"GLOF," said Jonathan. "Nineteen eighty-six."

GLOF, in the scientific vernacular, is short for glacial lake outburst flood, or as it is known in Icelandic *jökulhaup:* the sudden expulsion of meltwater from a glacier. This one had come in the summer of 1986 when a glacial lake breached its ice dam. Prior to 1986, big coihue groves had occupied much of this valley floor. What remains are busted-up tree stumps all tilted downstream in the direction of the flow. Mostly, there were no stumps at all. Forests had been bent over and buried under bright sand. Debris-strewn barrens topped it off, dotted with granite boulders the size of woodsheds.

A woman wearing practically a string top, dropped her pack, shouldered a video camera, and began panning us. She was perhaps one of the most beautiful women I have ever traveled with, the corners of her eyes marked with faint Egyptian back lashes that left a path of rumors and cell phone numbers tracking us across Patagonia. We had a two-camera setup. One was Denise Stilley, a twenty-four-year-old videographer from Flagstaff, Arizona. The other was Ed George, a grizzled filmmaker in his early sixties, wiry muscles and sun-cooked skin. He works for nature channels and has spent much of his life hoisting a camera over his shoulder on nearly every continent and in countless countries. They set up their angles. I started talking to the lens. Two producers, an odd couple of a tall, lithe mountaineer and her burly fireplug of a boyfriend, stood smiling and nodding behind the camera to keep my attention. I smacked a salt-and-pepper granite boulder as big as a car as if I were trying to sell the thing.

"This was shipped down here from miles upstream by one of

these floods. There are some powerful forces at work here, and whenever we get close to the ice, we see them amplify."

"Cut," the burly one called. His name was James Q Martin III, called, simply, Q, a wild-haired climbing photographer, tireless defender of wildlands, and now a filmmaker (this being his first film). "Technical difficulties," Q said. "Give us a minute. And why don't you put your hat back on?"

Ed lowered his camera, and they conferred. It seemed the light was wrong; a cloud needed to pass.

I put my hat back on, not sure if I liked this camera business, feeling like a game-show host. My job on the crew: scribe. While I waited, I pulled out a notebook and wrote:

Everything is about ice.
The shape of each mountain carved by this slow, relentless scour.
Glaciers dating back to the Pleistocene still exist, retreating like
*   hermit crabs into their shells.*
What the earth remembers: These mountains once looked like the
*   face of Antarctica, seamless wasteland. Now, this. It happens so*
*   fast, you catch your breath but barely.*

Over the last three million years, the earth's climate has been dominated by ice. We are still technically within an ice age, an age in which widespread glaciers still exist, relatively rare in the larger history of the planet. Ice has numerous times pushed from the poles toward the equator, squatting miles deep on top of places such as Chicago and London. These periods tend to last about 100,000 years and are separated by an equal number of warmer, wetter, and much shorter interglacial periods such as our own. In these interglacials, the ice retreats, though in the last few million years it has never melted entirely before turning around and starting up again. This climatic seesaw is fundamentally the result of the periodic, wandering tilt of the earth's axis. This tilt, about 23.5 degrees, is what gives us our annual seasons. It also gives the planet a much larger multi-thousand-year swing of seasons as our orbit wobbles slightly around the sun, the tilt creeping out and back by at least 1 extra degree. This difference of 1 degree is what drives the shift between glacial and interglacial periods. When the North Pole points more toward the sun, cold and hot cycles go to their extremes, winters become globally cooler, summers warmer. Warm summers melt the ice, and cold winters tend to be dry, failing to supply enough moisture to add to the ice mass. That sequence puts us where we are now: in an interglacial. When the axis straightens up by 1 degree, winters become warmer, summers cooler, and ice endures from season to season, resulting in a glacial period.

Deep ice-age conditions are generally colder and drier around the globe, leaving large portions of the Northern Hemisphere off-limits. Animals tend to be larger, the world more abundant with tusks and claws and big, stomping feet, not necessarily the best environment for skinny bipeds such as ourselves. There is a reason civilization and the rule of humanity waited until now to appear. These are our kinds of days. Everywhere I looked was a waterfall, mountains laced with white streamers, a landscape exposed and created by a changing climatic epoch. It makes you wonder what sorts of days lie ahead if the ice keeps melting and doesn't turn around as it has in the past.

For the last 65 million years the earth has had a Ping-Pong climate that changed dramatically on all timescales from multimillion-year cycles to decades-long collapses. There have been ice-free poles and continents half buried in glaciers, both flooding seas and oceanic lowstands. You could shrug these off as reasonable fluctuations on a complex planet, but that does not soften the blow when it happens. Fifty-five million years ago, global temperatures rose during a massive release of buried carbons. The earth lost its ice as CO_2 reached levels you would see now only if we burned all remaining fossil fuels. This Paleocene–Eocene thermal maximum was one of the faster worldwide transformations, though slower than what is being measured in our own brief time frame. Temperatures fifty-five million years ago boosted by about 10 degrees Fahrenheit. Extinctions spiked, especially in oceans where marine life took millions of years to recover. Meanwhile, other animals accelerated into the gap. On land, up went the orders Perissodactyla (zebras, horses, rhinoceroses), Artiodactyla (pigs, camels, deer, giraffes, sheep, goats), and Primates (monkeys, lemurs, gorillas, humans). Abrupt changes like this work out in the end, especially for those who make it through, but the moment itself can be trying, in fact catastrophic. Extinctions favor few.

"Okay, we're ready," said Q.

Yanked back from the waterfalls and tumbling seracs, I said, "Ready for what?"

That night my eyes flashed open to what sounded like a volley of artillery fire. The ground boomed beneath my back. There was nothing to see but blackness. I was camped in a snug one-man, and my eyes searched the dark roof of my tent; as I climbed out of a dream, my heart quickened. For a half second I was unsure of my location, my mind flipping through the deck of possibilities. Earthquake. Rock slide. Bomb.

Serac, I reminded myself.

It was not close enough to hit us, no bouldery blue fragments ripping through our tents like meteorites. We had set camp in a safe location where trees have grown from the retreat of glaciers, impacts long ago smoothed by Magellanic vegetation.

Echoes fired through the valley. They faded into more distant reports until all I could hear was the faint drone of waterfalls.

Awake now, I unzipped from my bag and danced barefoot out of my tent across darkened stream cobbles. Dew hugged the valley floor like fog. Everything was wet. Driftwood and grass dripped along the stony edge of a wide, mumbling stream. There was no moon tonight, sky sharp and bright with stars, the Milky Way brighter than I'd ever seen from up in the Northern Hemisphere.

A rumble came from the mountains, another serac falling valleys over. It felt like a great machine, geologic and climatic gears sharpening as they turned.

Had we been here in the 1970s, we would have been camped near the toe of a glacier. I had spoken with a gaucho who pastured horses and sheep up here since he was a kid, and he described a wall of ice hanging just above this forest where now there is a high, open valley.

Nearly all equatorial glaciers that were numerous a century ago are now missing. You can clearly say something big is happening, a warming, an acceleration, an ending of a time. The glaciers of Kilimanjaro—which Ernest Hemingway once described as "wide as all the world, great, high, and unbelievably white in the sun"—have all but turned to patches of hard-packed snow. No mincing with tape measures or comparing photographs is required to understand what is happening at this point in earth's history. You can see it. And if any doubt remains, gravity measurements taken by satellite show that since 2002 Chilean ice fields have been losing a net of about 30 cubic kilometers of ice every year. In mass, Greenland is losing around 240 cubic kilometers a year. Meanwhile, many scientists fear both the West Antarctic ice sheet and Greenland are destabilizing around the edges, as if we were witness to an epic, worldwide breakup.

Objects larger than anything humans can manufacture are coming apart. An iceberg four times the size of Manhattan separated from northern Greenland in the summer of 2010 and drifted south toward shipping lanes, threatening to decapitate offshore oil wells until it finally melted into the Gulf Stream. It was the largest iceberg seen in the Arctic since 1962, and it happened at the same time nearby Ellesmere Island was fracturing off some of its largest ice shelves.

The most studied breakup is the Larsen B ice shelf that once hugged the Antarctic Peninsula. With a surface area the size of Yosemite National Park, and a thickness of about seven hundred feet, Larsen B had been stable for at least twelve thousand years. In the summer of 2002 it came apart as if charges had been set, shattering over a few weeks into hundreds of pieces that soon floated away. That summer had been unusually warm, bringing with it prolonged melt that satellites recorded as ponds and lakes forming across the ice shelf's surface. All around the same time, these meltwater sources drained as if the ice sheet were opening from within. Three weeks later, it was gone, more than three thousand square kilometers of ice surface lost.

With this floating shelf removed, land glaciers that once ran onto it from land had nothing to stop them. They fell directly into the sea, now moving at eight times their original speed, dumping out twenty-seven cubic kilometers of floating ice every year. The supports came out and the ice fled.

I'd stop it here if I could. I have a particular fondness for this kind of earth, the salad days of the Holocene where we are somewhere between ice and not. No one can rightly say what lies beyond this point. What if ice completely melted? This is something that has not happened on earth for three million or possibly five million years —and some put it all the way back to thirty-five million years ago. A world without ice is unknown to us. We came into our own as a species across the lengths of several ice ages, incubated in Africa while climatic conditions revolved around glacial-interglacial swings.

The presence of ice keeps the planet relatively cool. Having a white surface means solar radiation is bounced back through the atmosphere and mostly into space. Other places like forests and asphalt-laden cities absorb radiation. At our current state, ice reflects a little more than 30 percent of incoming sunlight whose warmth might otherwise be captured by boreal forests like those of Patagonia. Henry Pollack, a professor emeritus of geophysics at the University of Michigan, worries that ice loss will magnify itself and the race to the finish could be faster than we imagine, putting us rapidly on a new kind of earth, a hothouse few are adapted for. He says, "During the last half century, radiation output from the sun itself diminished slightly, one of the reasons why scientists don't think the sun is controlling the climate system right now. It looks like human factors are driving this." He was speaking mainly about us topping off the atmosphere with insulating gases steadily inching up average global temperatures. As reflection surfaces shrink and solar radiation is more readily absorbed, the planet warms even more, causing ice to melt faster. "It's a very clear feedback cycle," Pollack said.

I asked him how long before the world's ice is entirely gone. He said, "Losing all the ice in the world? I think sometime between a thousand and ten thousand years encompasses most probabilities."

"A thousand years?" I asked, surprised at his shortest estimate. "I was thinking we had more time."

Considering that civilization was going full swing a thousand years ago, cathedrals popping up in Europe, China exploring distant seas, ten centuries into the future is not terribly far away.

"If warming accelerates, yes," Pollack said. "To be honest, we don't have a lot of experience with what happens as you begin to lose the big ice sheets of Antarctica and Greenland. You don't have to melt it all; you just have to let it slide off into the oceans. It could be that the final slide-off of really big chunks goes quickly."

Such a scenario is debated among scientists, some saying such a slide-off is not possible considering the friction between the underside of a glacier and the bedrock. But however the exact collapse happens, it can certainly pick up speed. Antarctica and Greenland are where most of the world's ice resides in masses that were for a long time thought too big to fail. That notion is now being challenged. Konrad Steffen, a leading climatologist and ice researcher, told me that when he first started studying in the 1970s and 1980s, nobody really cared about these massive ice sheets. "Greenland and Antarctica are very remote and were considered to be big ice boxes that responded not very fast to climate change. We never developed a mechanism to observe them until we had satellites and lasers. Now we see some surfaces lowering up to fifty meters per year." In case I did not hear him, he repeated himself. "Fifty, five–zero meters per year." That is a vertical drop of about 150 feet, far more than we were seeing in Patagonia.

What surprised Steffen was not the deflation of the ice sheets but their movement. "We thought they barely moved at all," he said. Steffen and his colleagues discovered that these ice masses, which are in places a mile or two thick, respond to even short weather events. Hot days can get them moving, whereas it was believed that they were going nowhere at all. Where are they going? To the sea, where, as Pollack said, they spill in piece by piece. The world's centralized ice masses are moving away from their centers, spreading out, and dropping off at the edges. In cooler or even wetter times, that ice would be replaced at the source. In most locations, that is not happening.

Some climate scientists, such as Steffen's colleague and vocal bellwether Jim Hansen, believe we have entered a stage of runaway ice loss, the big slide Pollack said could be coming, which would be a sign that the earth has taken a sharp and irreversible turn. Hansen describes glaciers surrounding the world's ice sheets as buttresses to a cathedral, and as glaciers decay, the massive interior ice sheets come spilling out because they no longer have exterior support. In Hansen's view, the cause is primarily anthropogenic, carbons released by gigatons into the atmosphere, wrapping around the planet like a shawl. Both Steffen and Pollack, as well as nearly every reputable scientist in the world, agree with Hansen on this matter. Humans are detectably warming the earth. Only Hansen thinks we've gone too far. As of 2005, our planet was storing nearly one watt per square meter more solar radiation than it was releasing, large by even earth history standards. If such an energy imbalance had existed over the last ten thousand years, all the earth's ice would have melted long ago. So Hansen, as well as many activists and scientists, believes we have come to a different state of being on earth.

Hansen's picture is of a rapid collapse, ice loss exponentially ramping up. Steffen won't go that far, though. He told me he is alarmed by the loss he is witnessing, but he is not convinced Hansen is entirely right. "For me, the good news is that there is not really a runaway effect, that we say it started now, and then we have a tipping point. That is where I disagree with Jim Hansen. We don't know what the tipping point is. Tipping point by definition means a place of no return to our current situation. I still think we can return."

Where they differ, Steffen believes natural systems are more resilient than Hansen is anticipating. Hansen believes we have overwhelmed any natural balance and are launching into a new and dangerous global state, a different world from the one we have known.

What they do agree on is that something large is happening on earth right now, something unprecedented in human history, and it is being driven by the disruption of global equilibrium. As Hansen is always quick to point out, we are increasing CO_2 levels in the atmosphere ten thousand times faster than they changed over the past 65 million years, and fossil sediments have even suggested changes similar to the end-Permian extinction 250 million years ago. Steffen does not deny this, but he believes we do not have enough data to say that we have entered a full downward spiral into a guaranteed iceless planet. After thirty years of working in the field on polar ice, Steffen says there is no question that the earth is warming, holding heat, losing ice. He has seen it with his own eyes. Like Hansen, he believes we are tampering with the upper limits of greenhouse gases, and on this biologically rich and sensitive planet it does not take much to change entire regimes, a change that would be devastating for almost every living thing, including us. Though Steffen believes it is not too late to cut back on emissions and turn climates around, he adds, "It is never too early."

So, how long do we have before all the ice is gone? Pollack put it between one thousand and ten thousand years. Steffen told me that Greenland probably has ten thousand years of ice at the current rate of loss, and Antarctica many more. But the point of a rapidly changing world is not some distant, singular event.

"You don't have to wait for the final ice to fall off," Pollack says, pointing out a direct relationship between melting ice and rising sea levels (the meltwater has to go somewhere). "One-meter sea level rise in my estimation is easily within grasp for this century, and it would displace 100 million people worldwide. That's where you're going to see it first." The effects of melting ice are far more widespread than sea levels, though. We are also losing the world's largest source of freshwater as it flows in increasing volumes to the sea. The glaciers of the Great Himalayas have been called Asia's Water Tower. The rivers they feed flow to some of our largest and longest-lived cities, which make up half of the world's population. During the dry season, which is often half the year, the flow of these rivers is mostly predicated on ice. That ice is melting so fast that right now river volumes are increasing from the source, but the future of such a trend is easy to predict. When ice is gone, water storage on the surface is over. Rivers run dry. And it will not happen gradually. The way ice melts, magnifying its own internal changes, scientists believe that Asia could turn on a dime. Studying how a warming climate would impact freshwater delivered from snow-dominated regions, the marine physicist Tim Barnett of Scripps Institution of Oceanography wrote, "The shortage, when it comes, will likely arrive much more abruptly in time; with water systems going from plenty to want in perhaps a few decades or less."

Can the earth really change that much, entering a new and culturally unprecedented episode? It has before. Considering the current state of change, there is no reason to think it won't.

In the morning, Jonathan was merciless. He brought me black tea double bagged just the way I like it. He set a steaming cup at my tent door. I groaned, realizing how early he must have woken, plenty of stars still out.

"Wake up, Craig," he said.

"Urf," I said, squinting for any light in the sky.

"We've got a long day."

"Smrg," I replied.

Jonathan had now taken the reins as our field producer. He quickly saw how disorganized we were and knew he had to crack a whip, ever so gently, to keep us going if we wanted to reach the sea anytime soon.

Today, horses stayed behind. We had a shot list to take care of, getting footage from around Nef Glacier, one of the bigger arms coming down this side of the Patagonia ice fields. We traveled in rough footing among cavernous heads of valleys, boulders polished by recent glaciation. By the time the sun was up and the mountains bolted with light, we were well along the high side of a pass, stopping to watch bigger seracs fall, hearing them as they cracked free, witnessing the silence of their flight as slabs quietly turned midair before hitting and exploding into what looked like comet dust. Ice was unfolding on a planet that behaves like a kinetic sculpture set into motion.

Teamed up with climbers and zestful adventure athletes wearing Day-Glo pro gear, I, the trip's frumpily dressed literary representative, was huffing to keep up. Even Ed George, that old bastard of a videographer, was a hundred yards ahead of me towing his camera over his shoulder.

Toward the head of the valley, we ascended gray backs of granite smoothed over by glaciers no longer here. The land was made of giant steps, like climbing over clouds. Carrying ropes and crampons, we followed the retreat of this lost glacier. You could date it as you walked. It had been at this spot thirty years ago, when subpolar deciduous trees were standing shoulder-high where they grew out of marsh grass, and fifteen years ago, when shrubs were stunted as tall as our waists, trees knee-high and gnarled. At the ten-year mark, there were only sprays of grass plugged into rock cracks, and then above that, where the glacier had dragged its way back, nothing had time to grow. We were approaching year zero.

Looking up, I saw wire-brained Timmy mounting a rock outcrop; when he reached the top, he whipped around and started flailing as if struck with electricity. With the energy of a hyper five-year-old, he cocked his body toward the sky, playing his trekking poles as if they were an electric guitar. When I came closer, climbing up to him like a supplicant to a guru, I paused for a breath. Timmy had an iPod plugged into his ears, the band Radiohead on repeat. As I passed him, he pulled the buds from his ears and deftly inserted them into mine. He dropped the player into my shirt pocket and took off up the rock as if he were a cat scrambling up a tree. I was left alone with a head suddenly full of music. This was anthropogenic change at its finest. I kept moving, finding the beat with my step, touching rock with the slap of my hand as the lead vocalist sang:

Everything,
Everything in its right place.

I climbed onto the head of the next-higher rock outcrop, found myself standing on it as if upon the dome of a mushroom. From there, I turned slowly, taking in the requisite 360-degree view. Waterfalls and high mountains closed in the back of our valley. They were overhung with glaciers, the tallest summits wearing white caps. Overdazzled by my surroundings, skull thumping with music, I found myself turning circles like a slow-motion Julie Andrews.

We crossed a low pass, its notch bald and glacier raked. Peering down the other side at year zero, I hit stop on the iPod. We were gathered at the edge, looking into a glacial wasteland. A hole the size of a football stadium had been gouged into the mountainside, where ice was left strewn and melting. It looked like a junkyard of blue-white bodies, and on the far side was the splintered face of a passing glacier that was still intact. The one we were following was gone, a tongue off the main glacier pulled out of the last valley over this pass and atrophied back to nothing. All that remained were fingers of glacier the size of Stonehenge sarsens leaning in various stages of breakdown over a murky gray pool at the bottom.

This was not what we had expected. There was supposed to be a two-hundred-foot deep meltwater lake backed up between the pass and the wall of the remaining glacier. The lake was gone.

"It was here two weeks ago," Jonathan said as he peered into the wreckage below. The lake must have forced a crack from the glacier behind it and rushed out. From the looks of its bathtub ring and the placement of once-floating ice, it had happened quickly, probably in one or two days. It must have been accompanied by a flood somewhere else, a GLOF charging into the wilderness downstream.

"This is how fast things are changing," Jonathan said.

It is easy to panic when you see how ice is retreating everywhere on the globe, floods coming loose, icebergs calving off. Even seasoned field researchers who have been watching this for decades are coming back amazed, saying glacial unloading has become unbelievable.

They describe in Greenland walls of ice hundreds of feet tall turning over one another as they plunge into the sea, the melt zone rising in elevation every year as margins break off faster and faster. Scientists who otherwise speak didactically with charts and numbers describe these events with their hands in the air. They have seen catastrophic releases with their own eyes, icebound fjords discharging all at once. One said that in the morning he was studying a glacial tongue filling the bottom of a fjord in Greenland. By evening all the ice was gone, fjord empty. "I've never even imagined a landscape change on that scale," he said, practically speechless.

Henry Pollack told me, "The warming we see now is unique in that it is rapid. It's taking place on timescales very different than those that govern the waxing and waning of ice ages."

Based on the climatic past, we are at a turning point. During the previous interglacial period, known as the Eemian (spanning between 130,000 and 114,000 years ago), conditions were similar to the Holocene, before entering a period of unsettled global climate. In an exhaustive study looking at evidence of catastrophic sea level rises from the Eemian interglacial, Paul Hearty at the University of North Carolina wrote, "We anticipate possible changes in our future that may include a series of erratic shifts in climate and oceanographic circulation, greater frequency and intensification of storms at subtropical latitudes, eventual [Greenland ice sheet] melt back, [West Antarctic ice sheet] ice collapse and rapid sea-level rise." Intense wildfires swept Europe during droughts that lasted for centuries, their ash carpeting the bottoms of lakes, leaving a perfectly timed record. Mega-droughts struck repeatedly as sea levels peaked and ice rapidly melted, and then the switch flipped. The Ice Age began and glaciers advanced across parts of continents and closed into ice sheets.

You could say we are there again, conditions mirroring the last millennia of the Eemian, climate and weather becoming increasingly erratic and unpredictable. Only this time, it's different. This time, we're here, and what may have been an already rocky journey into the next glacial period is going to be a completely different kind of instability. Many scientists believe a new ice age is off the table. We are ushering in massive landscape changes, river impoundments on a geologic scale. Through our exhaust and countless wildfires, we are overturning atmospheric chemistry, a phenomenon that is now creeping into the oceans, where nearly all indicators point to increasing warmth, CO_2 levels (which are correlated with radiation stored in the atmosphere) exceeding those of the Eemian. Some take our warming powers as a good sign. It keeps us out of an ice age. But it's not so simple. If you're not in an ice age and not in the mild Holocene, where are you?

I asked Pollack if he believes humans actually have the power to completely shift the global trajectory into some untried future. His response: "Can we add so much heat to the climate system so that the conditions that would have led us back into a glacial epoch are no longer adequate to take us there again? Yes, I believe so."

If true, I was traveling the collapsing edge of this change. I walked out to a rounded tip of rock. Across this enormous vista, everything was falling down into the bottom, boulders and pieces of ice the size of mansions tumbled toward what was once a blue-eyed lake and was now a shrunken, muddy pond. Timmy, not one to resist, was already running down the bathtub rings of this receded lake, his footwear punching through a steep, damp slurry of crushed rock. I saw the others getting out tripods, so I pulled out my journal and looked busy, descending off the pass behind Timmy.

Streams dribbled from cracks and hillocks, clear water running down steep, barren gullies glittering in sunshine. I walked through fragments of ice large enough to crush houses. The biggest I came upon was melting from the inside out. Fissures were cracked wide enough to walk through its low rooms, ceilings nippled with meltwater, the ground a gray slush. I glided fingertips across the surface. It felt like cold, wet marble.

Pop, said the glacier a couple thousand feet away. As I came out of the ice, I looked up at a facade a few stories tall breaking free and plunging a hundred feet to the ground, where it burst into white dust. Fragments shot in all directions, violent as crashing planes. Two-second delay, then boom.

"Pow!" shouted Timmy.

I turned to see him slapping the air, celebrating the explosion. Timmy was watching in awe as the eruption settled.

He looked down the slope at me and said, "Hey, you wanna do something?"

I followed Timmy to a lonely piece of ice, a big gnome cap sitting on the gray tilt of the abandoned lakeshore. He tossed me a palm-sized video camera and said to hit the red button.

I followed him with the camera as he jumped onto the flank of a fifteen-foot-tall piece of ice. Getting a leg up on the side, he started moaning and grinding against it.

"Oh, this is an awesome ice cube!" he cried.

He kissed the thing, running his hands up its side in breathless adoration, and I felt as if I were filming wilderness porn. Maybe this would be like one of those old film reels that survives into the future, something we would look back on, imagining what the earth was once like, surprised by how freely we toyed with it.

Timmy let go, ran at the camera, and with theatrical glee uttered, "We could have the world's largest gin and tonic. All we need is a really big tumbler, a giant lime, and a shitload of gin."

With a cackle, he sprinted out of frame.

There was one among us I would call a true, early-twenty-first century environmentalist, a loving, heartbroken activist putting her life into saving the earth. She was kicking in steps ahead of me as we climbed a steep ice bridge toward the roof of one of the glaciers opposite the vanished lake. Her name was Chris Kassar, a biologist by training and a coproducer for our film. She was the tall mountaineer. Out of her pack hung a forlorn-looking stuffed Lorax doll from a Dr. Seuss kids' book about the tolls of environmental degradation and human greed. Chris took the creature everywhere with her. After plenty of drink one evening I had found her sitting on a boulder with it seated by her side. Every once in a while, she leaned over and said something to the little thing. When I got up close enough, I could hear what she was saying.

"It's going to be okay, I promise, it's going to be okay."

Read: *they aren't going to dam the rivers, ice won't disappear, the earth we know will remain.*

After the earthquake hit southern Chile a couple weeks earlier, Chris had told me that it was a sign of the earth's anger at the way we have been treating it. "She's getting revenge," Chris said.

Scientists would mostly scoff at the idea, but there may be some remote validity to it. Ice is so heavy that it can literally flatten mountains, shoving their bedrock back into the earth's mantle. When the ice leaves, the mountains rise again, slowly, as if coming up for air. The Andes are currently lifting at an inch and a half per year, eight times faster than the Himalayas, partly from the collision of tectonic plates and partly as rebound from retreating ice. As ice leaves, it changes the behavior of the earth's crust. Volcanoes are statistically six times more likely to erupt during deglaciations, exploding as weight is lifted. Tectonic faults that were held fast begin to slip, earthquakes increasing in number and intensity.

In a briar patch of nonlinear dynamics and multiple forcing mechanisms, you could not safely say that this earthquake, the sixth-largest event ever recorded by a seismograph, was caused by climate change or by humans. But Chris, being sort of a jack-biologist (by degree but not by trade), could say whatever she wanted, and she believed the earth was taking revenge.

The times Chris and I had talked about this, I told her that I disagreed. I do not see the planet as angry, or godlike. It resists, relents, creates, puts forth, takes back. A busy planet, yes. Powerful, reactive, easily provoked to change. Maybe I have too much science in me, but a *process* seems more deep-seated than a *deity*. Deities die. The earth and its echo-filled cycles do not.

Whatever the case, this would have been an unfortunate time for the earth to take any revenge as we climbed the short bridge and crossed onto the roof of the glacier. As I kicked in my last steps behind Chris, I began to hear the glacier itself. I had expected it to be silent, but on top it roared and hissed with meltwater. Stomping around in metal crampons, I looked for the source of the sound and realized it was coming from within. Rivers were being born underneath me, inside the ice.

The terrain on the roof of the glacier was heavily broken up by its own movement, a landscape of nothing but ice made into kettles and crevasses. Fins stood in the sun, and below were coves of blue-bottle refractions. It felt as if we had climbed onto another planet. Meltwater ran down every crack and hole, burbling into slender, skin-smooth slots, tumbling into bottomless pits. As we scattered, I listened from place to place, here a plaintive whisper, and there a sound like ship engines rumbling belowdecks. In the lens-flare brightness of a cloudless mid-afternoon, I began to notice a slight shimmer on the ice. On closer examination, it was heat actually rippling off the surface. I had never even heard of such a phenomenon before, would have thought it impossible, warmth on ice. Opposites were in contact, ice burning.

This is what ice looks like as it leaves the planet. Under the sun, it funnels and furrows, drawing meltwater into its heart rather than sending it over the edge. This forms a feedback loop as water carries outside temperatures in, infiltrating the hard, cold interior and inching it apart, creating more melt.

Jonathan walked us over to a drill hole left by a research crew he had led here a year earlier. Dropped into it were lengths of plastic pipe cut every meter for measurement, held together by a cord so they could be lifted out. Whenever he crosses through, Jonathan stops for a measurement, which he sends back to the researchers. He reached in and started pulling up pipe, counting segments as he went. When he reached the bottom, he said that he had checked it three months earlier and in that time the surface of the glacier had melted by thirty vertical feet. "That's ten feet per month," he said. "That's how much is melting. Ten feet per month across fifty square miles of this glacier. It's mind-boggling."

As if forgetting about the film, everyone scattered. Jonathan was shouting to keep us together. But we were gone in all directions.

Ducking out of sight, I found the mouth of a narrow, downward tubular chute, a moulin into which meltwater plunged. You could have pushed a Volkswagen into it. I knocked in my spikes and leaned out over a trekking pole to see waterfalls pinballing into a dim labyrinth of sound. Baby blue at first, the moulin dropped into a shadowy saturated sea, a color so agonizingly rich it seemed perilous. I wondered what lay down there, a yawning, backlit cavern or dark racing corridors that would leave your tangled, bone-broken body jammed deep inside. You could not help imagine slipping off this rim slick as a bathroom sink, body shot through snakes and curves until you become a specimen to be melted out in years to come. When bodies have melted out after falling into a glacial interior, as they frequently have in the Alps, they are physically elongated, stretched by the movement of ice, and they emerge taller out the other side.

Fiber-optic cables have been snaked down into moulins, and even explorers in dry suits have gone down, discovering narrow tunnels leading into cavernous chambers within the glaciers. Ice is transparent down to ten meters. Below that, the inside blacks out. One glacier researcher described to me descending on a rope with a headlamp shining through sculpted vertical spaces intersected by numerous passages, each emitting its own waterfall. It was a many-chambered palace inside.

I jumped across narrow, grumbling crevasses and crouched at the sucking sounds of pipe holes. I thought it must be absolutely honeycombed down there, the inside of the glacier crisscrossed with streams, rivers, and hidden meltwater lakes. This is what a retreating, GLOF-shattered glacier looks like, its interior growing more hollow by the day.

When we finally regrouped and knocked off some of our filming, Jonathan slapped his wrist, indicating we needed to keep our eye on the time, didn't want to be caught too far from camp come dark. I wanted to complain, *We walked hard to get here. Please, can't we have another hour?* I saw the lowering sun, too, and I hated to worry anyone, but when Jonathan turned his back, I slipped away. When he turned to look for me, someone else escaped. The camera crew disbanded for some last-minute B-roll, and pretty soon he was alone. Who could resist? Peering into blue holes or balancing along the tightropes of ice crests, we were kids in a topographic museum full of ice sculptures. We were living in a world that still had ice, vestiges of the Pleistocene still in place. How long would this last? Would my children be able to see this? Would the generation beyond them?

I looked down a crack and saw Denise below with her camera. She told me I should come in. When I got there, she had her camera steadied on a narrow, melt-sculpted canyon barely opened in front of her.

"Are you filming?"

"Shh," she said.

She was filming, yes, and it appeared to be a film about two motionless walls of ice that come together like hands nearly in prayer. Inside was a narrow cavern winding out of sight, the trickle of a stream moving down its center. After about a minute, she pulled the camera from her shoulder.

"Is that the most beautiful thing you've ever seen?" she asked.

"Vaginal," I said, not really meaning to—*did I say that out loud?*

It *was* vaginal. We were looking at the bottom of a crevasse melted out near the edge of the glacier, a place you would have once fallen into and died, no way out of its winding, moulin-leaden passages. Now it was an unfolded, smooth-walled passage.

"You going in?" she asked, gesturing at her camera.

I looked back at her. Me with my poor little day pack and Indiana Jones hat.

I entered, slow at first, dazzled by the blue within blue of its walls. Denise followed, trailing me with her camera. Smooth as soapstone, curved as bellies, the ice drew back into itself.

When I didn't say anything, Denise reminded me, "We're rolling."

"Oh."

I was supposed to talk about something. About the immensity of nature, the power of ice. And I did. I turned to the camera inside this space burrowed out by warmth of air and runoff. In front of twisting, tightening walls, I hooked hands around cusps of melting ice and told the camera how good it was to be here, which was the honest truth. This felt like a fulcrum on which the earth is tipping, the inside of the ice coming out. You could not have done this a century ago, when this would have been a fathomless crevasse. It has melted down finally to this formation. We had come for the last moment, the opening of the final onion skins. Enjoy it now, I said. It may never come again, at least not for any of our kind.

I glanced up as a camera lens pointed down at me. Q spanned the two walls with his crampons, shutter flying as he tracked my passage from over my head. I kept going, feeling like a worm as I wrapped through tight, short turns. I shook Denise, whose video camera was unable to fit through the final squeeze. Q twisted his torso until I was out of his frame, my metal steps scraping along the last pinch where I could no longer be seen by anyone on earth, even the satellites overhead losing me in the ice.

I had to inch my shoulders sideways and jam myself in. It should have felt claustrophobic, but instead I felt as if I were being cradled in the blue palm of the earth. I hung in this last fluted incision listening to the stream as it continued into subglacial tunnels, its garbled voice trailing away. Jonathan was shouting fruitlessly for us to regroup. I could hear him in the distance calling all his kids back together. But we weren't coming. I'd later hear that at that very moment, two from the crew had squirreled away to have sex. A punchy river guide and her burly, soft-spoken boyfriend had dropped into a crevasse where no one could see them. Packs were pulled off, buckles unsnapped, pants tugged down. They didn't have a lot of time. She planted her palms on the ice, bracing into him as he held her hips and entered her. Later, I had to ask, they both reported they did it while wearing crampons.

To get a perfectly unempirical, casual observer's take on how long the earth's ice will last under current retreat conditions, I went to a friend, a frequent traveler and soil scientist. For the last thirty years he has been flying back and forth over Greenland on business, where from his twenty-thousand-foot window in a passenger jet he has watched the ice mass of Greenland deflate and retract. Flabbergasted by what he has seen, he said, "Fifty years. That's all we have before all the ice is gone. I mean *all* of it."

It should be mentioned that this conversation was held at our local brew pub, both of us on our fourth beer.

I balked. Impossible. It could not happen in fifty years. Not in a hundred, not in a thousand. I didn't even think Pollack was right. Most of Greenland's ice is land based, so past a certain point of decay it will no longer keep spilling directly into the sea. It has to actually melt to disappear. Meanwhile, the majority of Antarctica's eastern ice sheet is currently stable, also sitting on land, not prone to the degradation caused by contact with the sea. Fifty years is simply too fast for all that ice to leave.

Slurring his words, even if he did have a Ph.D., my friend said it's going to surprise us all, the ice gone just like that, *whoosh*. His hand swept the air.

There is only so much sense you can talk into a drunk man. I made a compromise, told him I would cut most of the peer-reviewed projections in half just to be safe, taking into account that every new discovery puts another dent in climate models, so far making the future planet seem only hotter and more unstable. In that case, Greenland has at least centuries. The soil scientist was not taking that for a second as he muttered, "Fifty fucking years."

When I repeated this claim to Konrad Steffen, the revered climate scientist gave me a dubious look. I told Steffen this is what some intelligent people think. Steffen called fifty years impossible. Let's hope so. We all want the Holocene to last. Well, most of us. Steffen said, "Some nations say great, we want warmer temperatures. In Greenland if you make interviews, two-thirds of the people say climate change is good, which indirectly is true if they just look at economic value. The fisheries have increased with warmer water. Halibut fishing now is at never-seen heights. Exploration has increased because the ice is retreating and companies are coming, they are chartering all the planes. So, economically, fifty thousand people in Greenland, I'm just guessing that most of them who are not traditional hunters say warming is to their benefit. That's my standard joke to the Danes. Make sure you keep good relations with Greenland because that may be the place you have to emigrate once Denmark is underwater."

Of course, there are many people outside climatic disciplines—of my friend's caliber—who say this is not happening at all. To avoid even their mention would be, well, unscientific. Alan Carlin, a headlining global-warming skeptic, told me, "Pay no attention, it's all based on bad science." At the time of our interview, in 2009, Carlin was an economist working for the Environmental Protection Agency in Washington, D.C. The EPA quickly distanced itself from Carlin when he began publicly decrying global warming as a hoax. From his cramped office, and speaking as a private citizen, he told me that climate change is bunk and humans are not causing what scientists say we are causing. "This is going to be the scientific scandal of the century. Temperatures are going down, not up," he said, basing his conclusion on his own review of satellite data.

"I'm not at all concerned about global warming," Carlin continued. "If you're going to worry about something, I'd worry about global cooling. Ice ages are really hard. A little warming we can probably cope with, but if the ice sheets advance, there is very little we could do."

He was right about that—ice ages are one of the greater privations a planet can face—but evidence for an oncoming glacial period is extremely spotty. Even if it did arrive, glacial onsets are one of the slower advances the earth can experience, requiring thousands of years of wet winters and cool summers to build continental ice sheets. Melting and warming, on the other hand, can happen fast.

I asked Carlin about melting ice, and he said, flatly, "It's a misreading of the facts." He fell back on a handful of global glaciers that are currently growing, a point skeptics cannot say enough about. These few advancing glaciers are offered as proof against what is becoming thousands of other glaciers on the melt, some retreating by miles per year. Carlin also cited relative stability in the ice across eastern

Antarctica, even a recorded increase on the east side between 1992 and 2003 (caused, you will recall, by moisture deflected south by an Australian drought). I gave him the usual arguments in return, ice empirically shrinking from all but a few locations, even large ice sources losing insulation and collapsing not just at the edges but deep within. New shipping channels had opened across the Arctic as sea ice retreated toward the North Pole, summer ice shrinking further than has ever been seen in history. This has shortened sailing times, spurring companies to design new hulls for Arctic travel in the absence of solid ice. Carlin stuck to his guns, saying climate change is mostly media hype and a skewing of science by the scientists themselves.

I wanted to believe Carlin, I really did. If he is right, does that mean we can have our ice back now?

On the shot list was "Tuesday Colonia Glacier Terminus."

On schedule, we crossed a couple valleys and entered the drainage of the Rio Colonia. In its head stood a blue-backed dragon of a glacier. A lake had formed at its toe, and below was a lifeless wasteland of GLOF debris. Jonathan had carried in a rolled-up inflatable kayak for two, and I had the paddle strapped to my pack, sticking straight up like an antenna. At the lake's edge, a freshly made beach of rubble and sand, we pumped up the kayak to shuttle one passenger at a time to the glacier's terminus. Jonathan did the paddling, navigating through the carnage-strewn exit wound of an outburst flood. This one had hit only weeks ago, everything still a mess. Jonathan steered, leading me across this meltwater lake toward the glacier's miles-wide terminus. Shards and slivers of ice floated all around us. From about 1960 to 2008, this glacier had been quiet as far as floods. In 1985 it began its most rapid withdrawal, littering its retreat with rock-thrust moraines you have to climb over. But it had not yet released a peep of a flood. In 2008 that changed. A meltwater lake gathered miles up onto the glacier, and suddenly it emptied and came blasting out of the terminus. Downstream, it took out everything in its path, leaving boulders of ice and rock strewn like chessmen across a devastated valley floor. The flood had so much force and volume that when it intercepted the Rio Baker forty miles away, it actually turned the river's flow around. For days, the Baker, one of the largest rivers in Chile, flowed upstream for at least twenty miles, causing extensive flooding upstream. Six months later, this same glacier unleashed a second flood just as big. Three months after that came a third, followed a month and a half later by a fourth. The one gauging station that could tell the size of these outburst floods had been turned into a twisted mass of metal brace bolted into bedrock forty miles downstream of here. It has been rebuilt numerous times, raised higher each time, and always the flood takes it out. The GLOFs were expected to dwindle over time, running out of source water, but these appear to be growing, both in volume and in frequency. It is as if an artery were opening, the Colonia Glacier emptying itself as its roof collapses with every flood. Now, only a few years after the first flood event, they were coming almost once a month. In Peru, a country especially susceptible to GLOFs due to its swift melt conditions, a sixteen-hundred-foot-high column of ice broke off a glacier in 2010 and landed in an impounded meltwater lake below. The resulting flood launched a seventy-five-foot wave down the valley, hitting a chain of towns, where miraculously no one was killed. Other GLOFs in the region have killed at least thirty thousand people since 1941. Similar events have happened in the Himalayas, where engineers have been pondering ways of mechanically draining the lakes before they can break through.

Are we causing these? Are we killing glaciers with CO_2 emissions, ripping them apart with floods due to anthropogenic global warming? That is too simple a question at this stage in history. Glaciers have been melting since the warming trend that followed the last glacial maximum twenty thousand years ago. The GLOFs you would have experienced at the end of the Ice Age make these look like trickles.

If you wanted to see a truly spectacular melt and a purging of ice, you would have come at the end of the Pleistocene around eighteen thousand years ago. Large portions of the earth, especially the Northern Hemisphere, were encased in ice until climatic tides turned toward the warmer world we know today. Ten percent of that original ice was gone within a couple centuries, a breakdown that must have been awesome to witness, something worthy of Hollywood. Enormous meltwater lakes collected across the Northern Hemisphere, dammed by receding glaciers. When those dams broke, floods up to fifteen hundred feet deep surged across parts of Siberia and the American Northwest, entire landscapes erased and re-formed. Most of eastern Washington now looks like a giant hydrology experiment, the Channeled Scablands made of hundreds of extinct waterfalls (some Niagara-sized), teardrop-shaped hills, and once-inexplicable drill holes into which you could drop neighborhoods, normal sediments up to two thousand feet thick swept away. A land area in excess of 1,500 square miles was erased. These floods came from suddenly drained meltwater lakes. They lasted a few days or even hours.

Even this Channeled Scablands flood was small compared with what came out of the largest meltwater impoundment of the time, Lake Agassiz, which stretched from North Dakota and Minnesota to Saskatchewan and Quebec, held back by the retreating Laurentide ice sheet. Twice the size of the current Caspian Sea, Lake Agassiz finally flooded out about eighty-four hundred years ago through Hudson Bay, where giant sand ripples and scour marks of icebergs on the floor of the bay suggest an absolutely cataclysmic series of events. So much cold freshwater was released into the North Atlantic that it appears to have altered ocean currents, perhaps stalling the Gulf Stream in what seems to have resulted in a global-cooling event that climaxed eighty-two hundred years ago.

What I was seeing was small stuff, the tail end of the ice. Flanks of fallen glacier lay half grounded in a shallow meltwater lake, surfaces brilliantly still. It was like gliding over a mirror, sun glinting through ice. As we passed close to some of the larger, grounded pieces, they sounded like chandeliers coming apart, ice tinkling and clicking inside itself. Riding as a passenger on the shuttle, I straddled my pack and glazed my fingers through bone-cold water. Below its chalky, gray surface you couldn't see half an inch. We were hissing through a suspension of digested mountain. I reached out and pushed smaller wayward ice out of the way.

"The earth can be a dark place," Jonathan said. "Beautiful, stellar, full of cycles, but brooding and powerful."

Jonathan talks that way, usually more cheerful, though.

He wasn't feeling completely comfortable with leading us back here, and he wanted to make me know that. He had come here before, but not with others. He didn't mind exploring the blowout with a friend, but having a whole crew was different. His guide senses were tingling.

My wife had expressed some concern before we began this leg of our trek. I had a conversation with her from an Internet café in Coihaique, and she had heard that GLOFs tend to be triggered by earthquakes. Considering southern Chile had just been hit with the largest quake in its history, she asked, "And you think this is prudent?"

I told her yes, mentioning that most of the large GLOF sources had blown out closer to the earthquake itself and had not had time to recharge. She kind of believed me. In fact, I kind of believed myself. It was such a beautiful, warm autumn day, who could imagine these glistening hulks of ice blowing out around us?

One by one we gathered on a piece of ice blocking the way. We carried the inflatable kayak over its sun-warmed surface, crunching as if we were stepping on wineglasses. A loud snap came from ahead. It sounded like the crack of a steel I beam from within the frazzled face of the glacier. We stopped and looked up. I shot Jonathan a question. He still seemed comfortable.

"You always hear those," he said.

Had he started shouting *Let's get the hell out of here* and spun in the other direction, we would have all started running.

Besides Jonathan, Q had witnessed an outburst flood, in fact in this very place. It happened at night while he was camped safely on a mountainside to my right. He later sent me an email from Chile, saying, "It started to break through the glacier wall at about 8:30 pm . . . I have usable photos till about 2 am . . . It's a lot to wrap your head around, I woke up and saw the glacier totally destroyed and the water level down with icebergs scattered randomly like hot wheels in a four-year-old's room." Q described watching "a berg the size of three skyscrapers split apart . . . sound of a train derailing . . . shot up maybe 100′+ above the original berg and then crashed into the glacier water . . . I felt like a helpless child . . . ."

The iceberg we were crossing was also about the size of three skyscrapers.

We stopped on the melt-sculpted surface for some filming in good, clear daylight. It was Timmy's turn this time. He put on crampons for show and tried to explain these floods for the layperson. He couldn't do it, though. Each take devolved, the word "GLOF" becoming a sexual by-product he showed by rolling up his pant leg and screeching at it dripping down his thigh. We doubled over laughing as Ed George lowered his camera, shaking his head. The producers said, "Okay, let's do it right this time."

Timmy finally shouted at the camera, "ROLLING VIOLENT DEATH JUGGERNAUTS!"

Take.

Jonathan came back from scouting ahead. "No farther than this," he said.

We followed him down to see the water ahead crowded with pieces of floating ice. Most were too big to push away. They did not leave enough room for the inflatable. Jonathan skimmed down to the edge. Reaching out with a trekking pole, he pushed one, and it rolled like a weightless boulder. The others bumped into one another.

"Nope, that's it," he said. "We stop here."

We were a few hundred yards away from the terminus, although it was hard to tell where exactly the glacier ended and where it was just fractured jumble still standing. Floods tear this place apart, some of the most destructive processes to ice. Calmer meltwater like you get every day actually lends to internal stability where the water uses dependable exit drains. GLOFs, on the other hand, blow the ice apart like bombs going off inside the glacier. The pipes and plumbing within explode, and the ice structure collapses. When the water recedes, these disintegrated passageways cave in. This is how you dismantle a glacier.

I wanted to go farther, but Jonathan was absolutely right. We were already standing on the borderline of foolishness. As I stared into the folded walls of the Colonia, Timmy crunched in his crampons beside me. I expected a brilliant snark, but he quietly studied the floating barrelheads that had stopped our passage.

"You know, I think I can do this," he said.

Do what? I was about to ask, when he took a running leap and was in the air. He hit the first piece of ice almost weightlessly, clipping it with the claw of one crampon. The piece rolled behind him like a log, but he was already gone, airborne for the next. He darted like a cat after a moth in the garden, leaping from one piece of floating ice to the next. He was beautiful, an animal. I envied him for that moment. He had wings.

Standing back and watching as he ran broad loops across the tumbling logs of ice, I felt a little middle-aged myself, glad all the same not to be risking life and limb. He was evidence to me that we humans have not forgotten how to move.

As if capering over wild hogs, Timmy came running back, aiming for Jonathan, who had positioned himself in front of the nearest large, floating piece of ice that Timmy could use as a springboard. Timmy took a full-stretch leap and planted a firm waffle-stomp crampon on solid ice. Jonathan spotted him with outspread arms, open hands. Perfect, no-touch landing. Timmy got shouts and applause from us all.

Even the Lorax would have clapped his furry little paws had he not been on Chris's back facing the other direction, seeing nothing but this tormented blast zone of ice.

"All right, it's late, let's start back," Jonathan said, relieved we didn't get any closer than this.

Aw, Dad.

We shuttled back through late-afternoon light, a golden mirror across ice-bobbed water. Back on the last iceberg and awaiting their turn on the inflatable, another couple from the crew crept out of sight and had sex. It must have been going around. Perhaps arriving at impending cataclysm arouses the urge to procreate, possibly the last great sex act in the world as we knew it.

Or, as she would later complain about the discomfort of having sex on ice, surprisingly not that great.

At the turquoise lake where we started on the Rio Baker, we switched to a flotilla of three oar-propelled rafts surrounded by a colorful flurry of kayaks. We left our horses behind for good. The remainder of our journey would be completed by river, floating about a hundred miles to the sea on a corridor of pure meltwater. A brilliant and boggling color, the river slid through overhanging woods. The water was topaz with the visible depth of sapphire. It was the color of earth, not dirt or rock, but the planet itself—sky, water, and mineral bound into a dazzling blue. You dip your oar blade into it and the glacier water seems to brighten.

Where gorges tightened into rapids too wild for our rafts, we portaged, letting Timmy and friends run it in kayaks, their bright Kevlar flashing as they went airborne down falls. The following chain of rapids was too much even for Timmy, the river drawn tight between metamorphic walls into a violent passage.

I asked him if he was going to run it, and he laughed, then shrieked, "What, do you think I'm a fucking idiot?"

We sent down one of our younger hotshots, a man in a kayak dwarfed by enormous rollers and colliding waves jammed into a crack in the earth. Watching him pound and slide through the bottom of the gorge a couple hundred feet below, I was glad not to be him. My breath stopped as he nearly rolled, then righted himself for the next shoulder-slamming wave. Mistakes in this kind of water are swiftly fatal. But he actually did it with grace and received full applause at the end.

I sometimes find myself clapping for my own species. What we are capable of is astonishing, like circus performers on the face of this planet. What else, I wonder, can we do beautifully, can we do right?

Down the other side of the last gorge, the river opened wide; we loaded back into the current, sweeping oars into water. On the front of my raft the Lorax doll was strapped like a sacrifice. After traveling through glaciers and forests, the poor creature was looking battered and more forlorn than ever. His own expression made me sad. I rowed his raft downstream, pointing him toward incredible views of ice-locked mountains. *Can you see now what a beautiful and perilous world this is? Look at it, Lorax. Even if this place were dammed, if roads were blasted into its walls, I would not let you turn away. I would make you stay and watch. Our world is not yet lost.*

None of this did I say to Chris, who lounged on a raft tube, her runner's legs tanning in the Patagonian sun, one of a crew of beautiful people. I kept turning oars through water, moving us on. That night we camped on the right-side shore, building a big fire out of deadwood dragged out of the forest. In a soft and intermittent rain with pants rolled up from moving in and out of the water, we were a bunch of Huck Finns in the wilderness. The meat we carried was not refrigerated. Jonathan hung the slick red slab of local beef in a tree, bathing it with wood smoke to keep it fresh another day. He cut off pieces and spiced them from a kit he carried, cooking them sizzling on a grease-black griddle.

The fire burned down as Denise began singing us a song in Spanish, a sweet, long lullaby. She had told me this place is magical, that spirits dwell in such landscapes. When she found me scribbling like a madman in my journal, she reminded me to be quiet, to pause. She told me a secret one night. She said she sees tiny windows and doorways in the forest, and once when she was a child, she saw faint lights inside of each. She believed the natural world to be magic.

"There is more here than us," she said.

In the morning, I stood at the river's edge beside Denise, both of us in raincoats quietly staring downstream. She had the camera focused on circles of raindrops touching the river's surface. A storm had gathered, settling its clouds into the mountains. The river moved inexorably underneath it. This is what becomes of ice. It melts and flows downstream to become part of the world's cycles. We were riding its wild knots and eddies, carried by the memory of glaciers. I looked over. Denise had turned off the camera some time ago and was just staring.

Nearer the ragged, sea-warmed coast, we moved into rain forest. In the coming days of rapids and waterfalls, the thickening woods on either side of the river became dominated by canopied deciduous trees. I knew most of the trees as Antarctic beech. Grown like the heads of green clouds, it is a relict species of the genus *Nothofagus* that once lived in Antarctica until crowded out by polar ice. The tree still exists in the southernmost temperate lands, Chile, Australia, New Zealand, and New Caledonia, its distribution telling a story of landmasses once pressed together 184 million years ago. That was the supercontinent Gondwanaland: Antarctica was mashed into South America, which pressed its cheek seamlessly into Australia, with New Zealand tagged onto the side.

Even then, the earth was dipping in and out of ice ages. Evidence from sediment deposits suggests that at times 500 million years ago Gondwanan ice spread as widely as it did during our own latest ice age. As the supercontinent moved and rotated on the face of the planet, ice migrated from what is now central and northern Africa to adjoining northern Brazil, continuing west to southern Brazil, into southern Africa, Bolivia, and northern Argentina. Over millions of years, these trees swiveled around these shifting glacial centers. Finally, they ended up here.

Finally.

As if this were somehow the end.

Day by day, clouds thickened until we were deep into temperate rain forest, two hundred inches of precipitation per year coming in off the sea. Mangy woods crowded the two distant shores around us. Six days down the Baker, a wet headwind blew against us. We traded oars and wore ourselves out turn by turn. Kayaks fished between our big, slow rafts until the lack of rapids bored them and they pulled their kayaks up on our decks and helped row. We tracked currents and ferried backward, pulling weight with our backs, our flotilla rounding forested bends. Ferns hung over the water, shaded by *nalca* leaves as big as blankets. This close to the Pacific, a single cloud layer decapitated the mountains. From under it fell countless waterfalls, so many they looked like brushstrokes on Japanese silk.

Oars made rhythms, circles grabbing circles, pushing us ahead. Mountains opened wide as other rivers entered the Baker from every side, flow added to flow. Those coming from the right originated at the Northern Patagonian Ice Field, those from the left at the Southern Patagonian Ice Field. The two largest remaining nonpolar ice masses in the world were blending into a milky-gray body that carried us through wind-whipped fjordal drizzle. We put up collars, pulled hoods, and kept rowing.

In the last miles of the Baker, the river divided into languid backwater bends. We followed a channel so narrow I kept scraping my oars on shore, oar blades brushing through ferns, grinding across sandbars. I stowed the oars and let the raft pass over sunken trees, current carrying me along. After a mile, the channel opened onto a large, chalky bay where I struggled to make headway into a choppy wind. Searching for the current, I spun my raft around and put my back into it. There was no current, I realized. The river was gone, and I was now in the Pacific.

The other two rafts ahead of me were moving into open water. Timmy had hauled his kayak onto one and was sitting proudly on a pile of gear. "We have reached the ocean, people, we have reached the ocean," he announced. "Well, Craig is still trying to reach it."

He turned to me and shouted, "Come on, Craig, you can make it!"

I'm in the goddamned ocean, I wanted to shout at Timmy, but it took all my concentration to keep a ferry angle. The object was to get to open water, break through shore waves, and set a course into one of the fjords for the nearest town while not making a fool of myself on camera.

Ed sat in the bow of my boat panning my every move as I tried not to have us be blown back into shore. Salt taste and fine grit splashed over the tubes, water cloudy with glacial inflow. Laid down as ice thousands of years ago, the water was released. Molecules had let go their rigid grasp, turned smoothly into a river, and then flushed into the fingered sea off the Chilean coast.

I think back home to Colorado, where my rivers are relatively small, waterfalls comparatively few. It was once like this there. At the close of the last glacial period, the Cordilleran Ice Sheet retracted and the interlinked glaciers of the southern Rockies fell apart while massive melt-rivers tumbled through the mountains. Now the rivers have diminished and glaciers have been reduced to a handful of small white patches, much of the state now semiarid. You see a possible future for Patagonia if the course does not change. You see atmospheric Hadley cells climb the globe as CO_2 levels rise, a different kind of world beginning to open as the ice keeps fading.

Finally shaking free from shore, I rounded into a spreading fjord and green and turtle-backed islands. I brought up the end of our line of rafts. Eventually, we turned inland one after the next. A short rounding of points led to the small cypress port of Caleta Tortel, where colorful wood houses and *tiendas* were built inside an ancient glacial cut. We lined up raft by raft at the town's cypress dock, equipment untied and tossed hand to hand. Locals came down to see what was going on. Cleanly dressed elderly retired gauchos watched from higher landings.

At the dock, our adventure was reduced to its lesser parts, gear boxes and straps piled on land. Our *hospedaje* was up an Escher-like series of wooden stairs. Windows were framed with lacy curtains. Tapping a propane water heater, we took showers and walked around toweling off, brushing wet hair. That night, dinner was at a restaurant with napkins and table service. We were unfamiliar to one another, clean for the first time in weeks. Our meal was *caldillo de congrio*, sliced conger eel floating in a bowl with onions, potatoes, and carrots. After we ate, conversation stilled. Glasses of wine were set down, bottles of Austral Calafate drained. No one thought to bring a camera. This is what the end of a long and windy day completing weeks of a hydrologic circuit looks like.

One by one we stood from our table, handing out kisses and good nights. I zipped my coat and walked into a damp evening outside, steps slick in the dark. The small town was asleep, its toes dipped into the water. Cypress steps led down to the sound of water gulping against piers. It slid up bare rock and back down. Tortel's official elevation was three meters above sea level, which seemed like an act of faith. Yet who would not go along with faith on a damp, sweet night like this? Wooden boats tied off rocked on gentle waves, clanking and creaking from one side to the next. I leaned on fog-slicked railing and looked into the dark.

As it touched my face, I could smell the wind. It was wet, borne across thousands of miles of the Pacific before hitting land for the first time. As it rose to the land, it let its precipitation fall. A brief rain came on this wind, and as it passed, I followed the sound through the wooden maze of Tortel. It swept higher and flustered up through the trees. Circles wound within circles. On tendrils and curls, the wind rose into mountains, where rain turned to snow and snow to ice as if it could not help reciting its old and familiar song.

3

Seas Rise

Stroke by
stroke my
body remembers that life and cries for
the lost parts of itself.

—Mary Oliver

BERING SEA, ALASKA

ON A SALT-SPRAYED KNOT along the north coast of St. Lawrence Island, a gray-haired grandmother huddled behind a rock pile trying to get out of the wind, not doing a very good job of it. The sea was as dark as the skin of a cucumber. It thudded and crashed against shore where we had taken shelter from a storm howling out of the Arctic.

"This is not summer," she shouted, incredulous, but not at me.

The island is a treeless, ferret-shaped landmass ninety miles east to west, and ten to twenty north to south. The Bering Land Bridge was once here. Now the land bridge is gone. This is what remains.

The woman cowering behind the rock shelter like a troll was my mother. She fumbled with a package of peanut-butter crackers she bought in the nearby Yup'ik village of Savoonga. She looked as if she were freezing, eyes tearing, mascara starting to run. Silver drops of snot dangled at the end of her red nose, wind blowing them off now and then. It wasn't that cold, really. Barely south of the Arctic Circle, we were on the north shore of St. Lawrence Island in early July. This *was* summer.

This is where the first people are believed to have stepped into North America around twenty thousand years ago. They came across Ice Age tundra with dogs and spears. This was no meager bridge at the time, no narrow causeway connecting the earth's hemispheres. It was a subcontinent all its own, Siberia blending into Alaska, where standing here in the middle you would have been as landlocked as in Boise, Idaho. Heading north from our rock shelter, the nearest coastline would have been five hundred miles away and you would have been looking across green-grass steppe, mammoths grazing among willow shrubs, great-horned herds on the move. It was one of the few Arctic landmasses not under ice, the lost subcontinent of Beringia. Had you told people back then what would become of this place, that water would have closed in making this an island, they would have thought you mad. There couldn't be enough water in the world to do this. It would have been an act of legend.

But it happened.

St. Lawrence Island is not as far south as the Aleutian Islands and not as far north as the fifty-mile gap where Russia and Alaska nearly kiss. Dormant gray volcanoes dip down to tundra-covered capes and bays where I periscoped my head up to see if anything was moving. Dark blizzards of murres and auklets swept out to sea and back, their shoal-like schools tracing the wind over our heads. I was looking for polar bears. They drift in with spring ice and become stuck on this island for summer. I scanned for a flick of white anywhere, but saw only birds. The last polar bear spotted near this east side of the rocky headland was about a month ago. Another had come within sight of the village at about the same time, spotted through the window at the grocery store where a man quickly dispatched it with a rifle. That bear was now a greasy pelt curing on a wood frame behind somebody's grandmother's house in Savoonga, about two miles from here.

Watching my mom work at the package of crackers, I took it from her, pinched open the plastic, and handed it back. She ate in silence. Short and with a sturdy frame, she's slender in the fingers, bony and chapped around the knuckles from her trade as a woodworker. Her hands have always grown cold easily.

"Give them to me," I said.

She put down the crackers and gave me her hands. When she was born, she was three months premature, kept alive by a lightbulb in a chicken incubator. I think because of that, she has always had a remarkable appreciation for warm sunlight on her face. She was getting none of that. I sheltered her hands against my chest, breathing onto them to keep them warm. Two and a half fingers on her left hand were cut off, the result of an accident late one night with a table saw while she was building a piece of furniture. The nubs that remained tended to be especially sensitive. Since I was a kid I warmed her hands this way, her eyes tearing in blowing snow or on high summits, whatever adventure she used to drag me off to. While we sat face-to-face, neither of us talked. I looked down at her hands and listened to the powerful boom of waves and the hiss of wind across rock. Seabirds flew overhead by the hundreds, seemingly unfazed by the storm, loud with chattering screams and whistling wings.

We'd been asked not to travel more than three miles from Savoonga. People in the village had been politely dumbfounded by the two of us, a mother and son up from the Lower 48 on an unlikely holiday, one a writer, the other his mother, in her mid-sixties, with curly-frizzled gray hair. The rock stack we crouched behind was an old hunting blind. Its poky, moss-bound floor was littered with rust-headed shotgun shells, and it smelled like a rookery, which it was. Murre chicks nested deep between cracks, and whenever one of us moved, the unseen chicks squawked and piped for partly digested krill. It was a little disconcerting, as if we were crushing the chicks, but they were safe down there. Even foxes couldn't get them out of their rock cracks.

I glanced out of our shelter where arcs of sea waves rolled in white and crashing. We'd been on the island for a few days and the weather had not changed. Yup'ik Eskimos whose lineage on this island goes back two thousand years had been telling us that seasons had been practically reversing, summer and winter turned upside down like nothing anyone could remember. They spoke of so much change, from shorelines to hunting to winter ice pack, that you'd think the world was being re-formed. But no one seemed especially alarmed, their statements relegated to Native nonchalance often mistaken for nobility. It was just the way people were, shrugging at the circumstance, wondering what could be done to change it.

Rough seas are nothing new here. Storms in the Bering Strait are notoriously violent and frequent. My friend Devin Vaughan, who walked the Mexican dunes with me, used to work on an aircraft carrier in the late 1980s, and he said that even on such a huge ship, when passing between Alaska and Siberia, veteran sailors would be vomiting. Rogue waves hit the ship so hard toilets on board blew like geysers. This roughness is partly a function of the sea being shallow. On average, the Bering Sea is only 100 feet deep, negligible compared with the Pacific with an average depth of 14,000 feet or the Atlantic at an average 12,800 feet. This water is practically more a veneer than a sea, an epicontinental water body that turns the space between Siberia and Alaska into a wave-amplifying slosh pan.

It didn't take much to imagine this sea gone. I drained it back to the Ice Age. I watched it shrink as a dry inland sun warmed us in this rock stack. Back to sea-level minimum, 340 feet below now, water rolled back to the horizon until there was no more and only land stood before us. This is what it was like here. The land was steppe, generally flat, and treeless as it is now. You would have seen woolly mammoth, bison, horses, camels, high-shouldered dire wolves, and a variety of rabbits, voles, and glittering little summer butterflies. Researchers drilling into the mucky floors of the Bering Sea have found peat, pollen, and terrestrial sediment, evidence of dry land now under a hundred feet of water.

"You're writing about the end of the world?" my mom asked as I rubbed her hands. "That's what you're working on now?"

"Kind of," I said, eyes still scanning the foam-waved rollers, seeing them gone.

"I don't like thinking about that sort of thing," she said. "Do you really think it's going to happen?"

"What?"

"The end of the world."

"It's always happening."

She frowned at me, a face of many words, most saying, *Don't get smart with me.*

A sheet of a few hundred murres flew over us, and the chicks below gabbled, their peeps and rustling little wings calling attention. The entire headland sounded like this, the ground a sort of intermittent clamor.

The globe used to be laced with land bridges. But for a few short swims, you could have walked all the way from Australia to North America during the Pleistocene, crossing the lowlands of what is now Micronesia, up Asia, and across to Alaska. Records of dead, drowned corals and submerged shorelines around the world stand as testament to the rise. An ongoing National Science Foundation project has been mapping underwater geographies off the west coast of Florida, tracking the courses of rivers that once flowed. At the confluences of these now-submerged rivers, researchers out of the University of Florida have discovered ground stone arranged into irrefutable evidence of human encampments dating back to around ten thousand years. Around the world you find the remains of settlements underwater, almost every coast rimmed by swallowed archaeological sites. The Geological Survey of Canada has used high-resolution digital terrain imaging and seafloor sampling to record underwater landscapes off the coast of British Columbia. In five hundred feet of water it found tree stumps still sticking straight up where forests used to stand, and it dredged up from one of these ancient shorelines a stone tool encrusted with barnacles and tentacled bryozoans, evidence that humans had lived in places that are now miles out to sea. Great Britain is surrounded by underwater Bronze Age towns, India has its sunken coastal temples, the Mediterranean its stone sculptures and drowned port cities. Half of ancient Alexandria is now submerged, its pillars standing majestically in shadowy blue depths. Almost every coast is an Atlantis.

The noose has been closing on us for many thousands of years. There is no reason to believe sea levels will miraculously stop here. Looking across the waves, I searched for humans moving with spears and dogs. Berries were being gathered in baskets, roots dug with carved bone tools. You'd see lines of smoke blowing from summer villages, skins stretched over wood frames. There would have been laughter among children, or a woman standing alone, combing her hair, studying the eternal skyline.

"I'm going to keep walking along the coast," I said.

My mom nodded and said, "You can take my crackers."

"You keep the crackers."

"I'll start back for the village."

"We'll meet on the road."

She nodded again. If I were a good son, I would walk her back the two miles, but she is not someone I worry about, quite capable on her own. After I got back from my trip to Patagonia, she was so jealous of the adventure that a year later she flew to South America and retraced our entire trip from glacier to sea. With much worse weather than I had, she packed over glacial moraines, stomped across ice, and portaged most of the rapids we ran. She came back haggard and happy. "I have never been so cold and so tired in my whole life" was the first thing she said to me on the phone. "I had blisters where I never had blisters before. I fell in love with everything."

I don't worry about her.

The village council in Savoonga had let out its coffee-break meeting room to us for forty dollars a night, complete with a couple cots, a flaky old conference table, and more desk chairs than we knew what to do with. On a day like this she would be comfortable there, and she would have visitors. She loves visitors. If you sit around the conference room too long, women start coming by with fresh-baked bread and cookies wrapped in aluminum foil. Some would sit and talk politely about the well-being of their families. Men would bring carvings to sell, and when you were finally out of money and could not buy anymore, they would sit smiling, nodding, telling stories of what it is like to converge on a bowhead whale from the bow of a small skiff in a dangerous sea, harpoon raised.

My mom zipped up her day pack, the same dumpy old one she'd been using for half her life.

"Watch out for polar bears," I half warned her.

"Thanks," she half joked.

During the last glacial maximum of the Ice Age, this region was the place to be, the heartland of Beringia, the center of the land bridge. When much of the Northern Hemisphere was encased in continental ice sheets, this was open ground, kept dry by an inland draft coming off the Pacific. Life was equally diverse back then, only it was land based with hardly a fish or river to be seen. With the sea up, conditions have changed. All that remains of the megafauna are reindeer herds managed by Yup'iks, the occasional polar bear, and, if you want to stretch it this far, a common wolf. Everything is different now.

The shifting of sea levels is far more than a superficial change, a mere rearrangement of shorelines. When the Bering Land Bridge was finally covered over about 9,500 years ago, the labyrinthine water clock of the earth shifted. The Pacific was let through into the Arctic Ocean, and hemispheric waters mixed. One side of the world, the Pacific, has far less salt in its water than the other, the Atlantic. The slightly fresher, less dense Pacific now moves up and over the pole and comes down on the other side, rearranging currents. Currents both warm and cold, dense and light, channel through the intricate straits and passages of Baffin Island, Newfoundland, and Greenland. When these currents switch, climates change.

Aixue Hu, a climate scientist for the National Center for Atmospheric Research in Boulder, Colorado, recently ran a computer model to see what might happen when the Bering Strait opens and closes. Hu believes this mixing of water may be what turns ice ages on and off. While the earth's axis plays a major role in creating glacial and interglacial periods, the earth itself fine-tunes the process, and scientists have long been looking for what actually governs the handoff from one major equilibrium to another.

According to Hu's model, when the Bering Strait opens and lets the Pacific through, the North Atlantic eventually becomes less saline, its water less dense, which slows the warm, low-density flow of the Gulf Stream. Once this warm Gulf water no longer circulates into the Arctic, the top of the Northern Hemisphere grows cold and switches to a glacial period. Water once more locks up into ice. Sea levels drop. The Bering Strait closes, stemmed by the land bridge, dividing oceans. At that point, the North Atlantic returns to more saline conditions, and the density difference begins to draw up the lighter Gulf Stream until the Arctic warms, ice melts, seas rise, and we enter the next interglacial period. Ice grows, seas drop, and the cycle goes on.

Whether Hu's hypothesis proves out, it is at least a window into the entangled functions of this planet. When his findings came out, Hu said, "The global climate is sensitive to impacts that may seem minor. Even small processes, if they are in the right location, can amplify changes in climate around the world."

I had come to stroll in a place where the world changes, a clutch of mountains threading this swirling sea. This is one of the earth's switches where a weight shifts from one side to the other. Water whipped off the waves as if it couldn't stay put. Mist streaked across the side of my rain hood. In one direction was a mantle of treeless tundra gently rising to the cloud-cut, snow-laced Kookooligit Mountains, and on the other water hammered at the island like a man on an anvil. As I looked out from a spit of whale bones and rocks, the twenty-foot rollers seemed well over my head, as if about to swamp the island. Vertigo. Optical illusion.

We do not live in a particularly impressive period in history for watching sea levels rise. Go back to the swashbuckling waters that swamped the Bering Land Bridge starting with the first upsurges about eighteen thousand years ago. It would have seemed a cataclysmic time, the rise of seas measured in feet per year rather than tenths of inches. Seas rise more slowly along steep coasts (California, Dover), but quickly along shallow platforms (St. Lawrence Island, Florida). You couldn't have built coastal villages fast enough.

Continental ice shelves were collapsing, letting out enormous meltwater floods as big as Pleistocene Lake Agassiz, enough outflow to the sea to briefly chill the oceans, plunging the earth into erratic postglacial cold periods. These floods were felt around the world, with the Mediterranean rapidly overfilling and spilling through the Bosporus Strait into the Sea of Marmara where sandal-wearing pastoralists would have been overwhelmed, giving rise to the legend of Noah's Flood. This stage of rapid sea-level rise (paling our piddling sixteenths of inches) ended about seven thousand years ago. The volume of ice no longer existed to keep up the pace. After that, the rise was more gentle, falling into lulls like the one that began around the first century A.D. Estimated from shoreline benchmarks in Tasmania and the height of ancient Roman fish tanks, sea levels changed only slightly for another two thousand years after that, enough to swallow various islands and shorelines around the world, including the margins of St. Lawrence Island, but only by small amounts. This quiescence ended about two hundred years ago. Near the close of the eighteenth century, right around the time of the Industrial Revolution, tide charts began recording the upward surge that we are in today. To the best of scientific knowledge, sea-level rise accelerated through the twentieth century and is continuing to pick up the pace at the beginning of the twenty-first. Some scientists see this as our entry into a rapid, coast-destroying rise, other scientists say that recent, highly accurate data sets have not been collected for long enough to indicate any runaway trend. Either way, now or later, fast or slow, it's coming up.* Gravity measurements taken by the Gravity Recovery and

---

* One of the lone dissenting voices in the scientific community on the current sea-level rise is the retired Stockholm University oceanographer Nils-Axel Mörner. Well published and peer-reviewed, Mörner interprets tide charts and coastal markers, which he says show ups and downs but no significant recent trends, and he believes sea levels have hardly budged in forty-five hundred years. When in 2010 the president of the Maldives island nation, at extreme risk from potential rises, held a conference underwater in scuba gear to advertise his island's plea for assistance, Mörner wrote a terse letter to him, saying, "Your people ought not to have to suffer a constant claim that there is no future for them on their own islands. This terrible message is deeply inappropriate, since it is founded not upon reality but upon an imported concept, which lacks scientific justification and is thus untenable. There is simply no rational basis for it." Mörner's claims that sea level rise is not happening have been widely discredited by many scientists using gravity measurements and their own analysis of tide charts to show that average sea levels are, indeed, coming up.

Climate Experiment (GRACE) satellite, launched in 2002, reveal a steady increase in mass of the entire global ocean, so whether you believe in sea-level rise or not, the expanding mass tells the story.

I have a trepidatious regard for any body of water that I cannot see the other side of. Born and raised inland, I feel comfortable on the ground. Whenever I walk to the edge of the sea, the sensation of immensity and inaccessibility comes over me. In my mind, the world ends at the sea. Salty, oyster-gray clouds of spray billowed over tundra like some diaphanous marine creature losing form and disappearing before the next wave rushed in to take its place. It felt as if the sea were trying to climb onto the island in every gust.

I looked out beyond wave-wet rocks and stranded whale bones with colorful oil bottles, thinking you would need paddles, waterproof skins, and hulls to survive past here. Even for its seething genetic diversity of krill, seal, and whale, it was not a hospitable-looking place. Simple wits might buy you a few minutes, but the water is cold.

I was told a story in Savoonga about Siberian neighbors who once paddled their umiaks across for a celebrated reunion. Being in U.S. waters right up against Russian waters during Soviet occupation, these visitors had strict requirements for their return. At the end of their visit, a storm came up, but they were compelled to leave for fear of being charged with defection and the retribution that would ensue. Waves rose and swallowed their boats. Islanders watched helplessly as every last one capsized into this cobalt sea. As the story was told to me, all were lost, and I imagined infants and children sinking like stones.

The time of dry land that you could walk across to get here and back is over. Driftwood has washed in from Siberia, tangled up with bright fishing buoys and old rope. I stepped over slick rocks, finding whatever else floated in from Oceania, the world's single, interconnected ocean: bottle tops from Polynesia, oilcans from California, and a rounded scrap of pressboard from a Japanese tsunami, I thought.

This was about the three-mile mark, time to turn inland. Meadows of grass bent away from the wind. Savoonga could not be seen over oceanic waves of green, misted tundra, but I kept my proximity in mind, maintaining a three-mile line out of respect for what they asked. I had trouble giving the water my back, turning around to watch its white rollers receding behind me, as if they might do something sneaky.

There is no such thing as one definitive sea level, no one, single height at which world oceans stand. Levels change all the time, from waves to tides to warm upwellings and cold, salty currents that sink. A term you often hear from scientists is "eustatic sea level rise." That is, overall mean sea level, which is different from local mean sea level, which is a mess, all over the charts depending on when and where you are. The Indian Ocean in the vicinity of Zanzibar is incrementally lowering over time, while the nearby Bay of Bengal is seeing a measurable rise. The same is true for the Eastern Seaboard of North America and the southwest side of Great Britain, where it is actually an uphill journey across the sloped Atlantic from New York to London.

Think of the oceans as pans pouring back and forth into one another, driven through a gauntlet of continents and isthmuses with different points of influx from both freshwater and thermal expansion. Tides set up by the push and pull of sun and moon spin around the globe, and their friction against the earth's surface actually slows planetary rotation by about half a second per century, brake pads resisting the turn of the wheel. It is no minor thing that a quarter of the earth is covered with water. It is another one of our elements of perpetual change.

Carl Wunsch, a foremost paleo-oceanographer out of MIT, told me, "The ocean responds on all timescales. There are elements of ocean that react quickly. Turn the wind on, and you can have big waves within minutes. But say you want to set the whole ocean, say the whole North Atlantic, into motion by making the wind stronger, that's going to take decades to hundreds of years. It's misleading to say there is one timescale."

The behavior of oceans depends on where you are. Louisiana has been losing the majority of its coast by an average of twelve meters per year to erosion, while large parts of the Nigerian shoreline have seen a retreat of thirty meters per year, one of the fastest geomorphic processes on earth.

The ocean, Wunsch said, presents all sorts of priorities. From the vantage of centuries, we see a relatively gradual rise of inches, while it matters whether a hurricane hits at high or low tide, determining whether a coastal city is half plunged underwater as levees back up and breach or it's just a passing storm.

"Sea levels have been changing for thousands of years," Wunsch said. "Certainly going back to the last glacial period. So, that's a ten- to twenty-thousand-year timescale of change. But you know, sea level goes up and down twice a day by two meters. If you live in Japan, you had a tsunami that came up by many meters—though it didn't stay up—in a matter of minutes. It's really very difficult to say the ocean reacts only on long timescales or even on short timescales. You give me a time, and I will find some element of the ocean that responds in that time frame."

One of the major changes is temperature. Half of current sea-level rise is attributed to thermal expansion, ocean volumes increasing as water warms and spreads out. Sarah Purkey, an oceanographer at the University of Washington, recently synthesized deepwater data from around the world, finding that heat is measurably creeping below three thousand feet, expanding the undersides of oceans around the world. Purkey and co-author Gregory C. Johnson concluded that the deep ocean is warming as well as the upper ocean. Their findings indicate that warmth sinking into lower levels is causing oceans to heat 16 percent more than previously thought, a significant increase when trying to calculate future climates. Slow to warm and even slower to cool, oceans are the largest reservoirs of heat on the face of the planet. On land, weather is capricious and changes fast. In the sea, changes are slow. Even if we stopped global warming right now, it would take more than a thousand years for the heat to be reemitted to the atmosphere. Since the 1970s, heat content has increased in the upper levels of ocean water; one oceanographer said the amount of heat collected there between 1993 and 2008 is equal to the heat of

five hundred 100-watt lightbulbs per person on earth burning con-
tinuously, their heat sucked up and held by the water.

I spoke with Purkey about her findings, and she explained that
while the upper ocean can change rather quickly—for example, El
Niño Southern Oscillation is forced by dramatic changes in ocean
surface temperate that occur in less then a month—shifts down deep
take much longer. "If we turned off deep convection in the southern
ocean, the northern ocean would not even feel it until forty years
later," she said. "The amount of time it takes for water to actually
travel from the deep southern ocean to the deep north Pacific is
about a thousand years. But there is communication between these
two regions on much shorter timescales. We have seen a lot of warming in the deep southern ocean, and this warming will affect the
ocean's pressure field (similar to when you heat the air in a tire, it will
increase the pressure). This change in the pressure field is being felt
thoughout the ocean in about forty years, causing the deepest coldest
layer of the ocean to warm around the globe. The warming moves
through the ocean by waves, just like a surface wave you see on the
beach moving toward shore, except these waves are in the interior of
the ocean. Like a surface wave, it travels at a given speed. Because the
ocean is so big, it would take forty years for this wave to propagate
around the world, therefore the warming we see in places like the
north Pacific, were caused by a wave that started forty years ago."

If we boil things down to their most simplistic level, we may now
be seeing oceanic changes from what was happening forty years ago
in the 1960s, when human population was at three billion, less than
half what it is today, and when we barely even thought of sea level
rise. What we get forty years from today's nearly seven billion people
remains to be seen.

Purkey reminded me that there is nothing simple about the way
oceans behave; you cannot just assign a number of years and say that
is how long it takes for them to respond. Oceans behave fluidly, are
hard to pin down—a bit of chaos added to our fortune of being here
on earth.

Carl Wunsch has written that the earth "is capable of remarkable changes without human intervention." He admits that distinct human influence is difficult to tease out and separate from what is being seen in ocean changes. Wunsch thinks of the future in terms of probabilities, expecting sea levels to rise in a climate of anthropogenic warming, even more than natural variability might allow. "Catastrophically or not, I don't know," he said. "Nobody knows. Maybe we never will know. But would it not be sensible to take some precautions that are comparatively cheap at this point? It's a worst-case scenario, I don't even think it's likely, but if sea level came up by a meter in thirty years, you'd have a calamity. It would cost trillions of dollars to deal with, and you'd probably kill a lot of people. Well, it wouldn't cost trillions of dollars to take precautions now in terms of reducing CO_2 emissions, trying to do something about discouraging people from settling in low-lying areas, doing something about water supplies in coastal regions. Look at Bangladesh, where the population in low-lying areas has quadrupled over the last thirty, forty years. If somebody, some omniscient government, thirty or forty years ago had said this is a very bad thing, they might have prevented what could now be a truly catastrophic event."

Even if climates evened out or cooled slightly, you'd still be looking at another thousand years of rise based on thermal momentum alone, the sea holding on to whatever heat it gets. The question is not whether it will happen but how fast. Speed is everything. The average car, for instance, goes from zero to sixty in several seconds, which has little effect on the human body other than pushing you back into your seat. If a car went from zero to sixty in a tenth of a second, you'd be dead. In the same respect, if sea level rise happens slowly, as in inches per century as it has in the most recent past, it is a matter of gradual attrition. Our coastal cities lift their skirts and step back as old piers molder into the water and former waterfronts sink from memory. Faster than that, we call it devastation.

Speed is modeled and forecast but ultimately unknown. Many leading scientific predictions from the 1980s and 1990s are falling short of our current eustatic rise at half an inch per every several years and increasing. An acceleration, at least over the last ten years, has been detected. This matters because 40 percent of the human population on earth lives along a coast. Twenty of the world's thirty largest cities with ten million people each or more are on the water.

Set aside time for a moment, and look just at volume. How much more water is available to top off ocean basins? To reach sea level maximums our world hit in the previous interglacial 100,000 years ago, we have about twenty or thirty feet to go. If we go beyond that line, all ice in the world melting, including the poles, which hasn't happened for millions of years, it would be about another hundred feet on top of that, which would put nearly all of Florida under and pull the Gulf of Mexico up to Memphis. The Netherlands would drown, left as scattered islands, while China's largest cities would vanish beneath a hundred feet of water. I know, it sounds like an act of legend. Start talking that way, and people will think you're mad.

Shishmaref is an Inupiat village with ancestry dating back as far as the Yup'ik. Literally falling into the sea, Shishmaref has been disappearing house by house. It sits on a few square miles of an island off the north coast of western Alaska 220 miles from St. Lawrence Island. Barriers of all sorts have been put up to stem the loss. Sandbags and cabled blocks of concrete, each has failed. Some have even exacerbated the island's demise. It is not as simple as sea level rise, though. Shishmaref backs up what people from Savoonga told me: Things are changing faster than anyone remembers. The loss of permafrost is pandemic, and Shishmaref sits atop a permafrost island. As it melts, the island unfolds from the inside. Warmer winters and changing wind patterns have prevented shore ice from forming, which exposes the island even more to storms and rising seas. Buildings have been jacked up, moved inland, but they have little room to move, in some places a few hundred feet to the island's back shore.

St. Lawrence Island is higher and will last much longer, a refuge even in the highest sea. Hummocks were gnarled with lichens. Rusted oil barrels appeared now and then, some capsized, some standing like the Rapa Nui heads of Easter Island. Dripped with white streaks of raven droppings, they were markers for travel routes; I imagined spots of darkness appearing across a winter-white landscape, the road home. Half a mile in from the coast, the sound of the sea had washed into a single roar sometimes indistinguishable from the wind.

Something moved across the slope of the island. I picked out a four-wheeler riding down a hill a quarter mile away. It turned toward me, jouncing across tundra as if driving through a tossed salad. Wind-feathered grass dipped into swales where the four-wheeler finally pulled up with a growl. The driver flipped the key. He wore oil-stained coveralls, and we paused to study each other before saying a word. His engine puttered to silence. The middle-aged Yup'ik man with a weather-slicked face leaned on his steering rack.

"Thought you were broken down," he said.

"Just out walking," I said.

His eyes judged me. I was not dressed anything like him, with my layers of undershirts and cap. He was from the western linguistic groups of Siberian Yup'ik, people who had been here, or at least somewhere near here, for a couple thousand years. A northern-born culture, his bloodlines had known northern country for at least twenty thousand years. He had a shovel and a rifle strapped to his rack. I had a day pack with a water bottle and a sandwich my mom made. He gestured behind him, toward the mountains. "There are wolves out there," he said.

"Yeah, I heard."

"You don't worry about wolves?"

"No, I worry about them. I'm on my way back to the village."

"They got shoulders as tall as a Honda," he said, clapping a hand on his fender. Nobody in the village had a car. Just Hondas, four-wheelers. "Can bite right through steel," he insisted.

People know what to do with bears here, having had a long and continuous relationship with them. Wolves, on the other hand, had been extirpated from St. Lawrence Island centuries ago. These likely came across on winter sea ice, probably from Siberia, about ten years back when the ice was much thicker and more expansive. They followed the track as if the land bridge were still here, as if they had atavistic knowledge of the way to go, island hopping as they may have done thousands of years ago when the sea breached this place. So far, no one has been able to kill them. They hide in the Kookooligit Mountains and come down on stormy days, when they say you sometimes see them like ghosts in the fog.

I told him I would keep my eye out. He chewed on a thought for a moment, then grinned at me. He took off his glove to shake hands, introducing himself. After a short bit of talk, he lit a cigarette, leaned over his brake, and asked what exactly I was doing here. I told him about wanting to stand with my own two feet on the last piece of the Bering Land Bridge. This is where continents once met and where they split, a point that changes the earth.

"You got a land-crossing permit?" he asked, still grinning in a way that seemed both welcoming and circumspect. He was thinking maybe he could hire himself out as a guide. I was meat on his table, a carving knife, or a year's supply of Pop-Tarts from the cavernous village store.

"No, not yet."

He dragged on his cigarette, gestured toward the farther coast, horizons of tundra rolling over each other. "When you get a permit, I'll take you out there."

"I'm here with my mom. Can you take her, too?"

"That's your mom?" He looked back over his shoulder. He had obviously encountered her on the way.

"She's gonna get eaten by wolves," he said.

I've spent time on the sea, not much, but enough to get an idea what it's like. The clearest picture I have is with my mom. She was in her forties at the time, I was in my twenties, and we paddled the coast of Belize for several days together. In a pair of gear-heavy sea kayaks, tents and food jammed in around our feet, we hopped from island to island. I had a compass mounted on the deck in front of me, and in long reaches where you couldn't tell an island from a skim of cloud, I read tides and counted waves, feeling as if I were in a room full of clocks. How could you even concentrate, chimes going off, pendulums ticking, the sea in constant rhythmic motion? In a way, it strikes me like an ice landscape only faster and smoother. We had fishing poles rigged out the back, and every night we camped on a different island among bowlegged palm trees or fish-shaped spits of mangroves. Navigating between the farthest ones was a practice in infinite perception, eyes roaming the bright blue bifocal of sea and sky.

One day, we spotted a shack out on the water, a small Garifuna fishing settlement up on stilts. It looked like a stork two miles out. We changed course for it on a midway point between islands. I expected there to be more, but it was one shack standing in shallow water, no land to be seen. It must have been on a reef, maybe a former island stripped out by a hurricane. Paddling in closer, I could see the shape of the island about two feet under water, coral cleared for sand. A few excited, bare-chested kids were on a sort of dock attached to the structure. The kids were animated in the sun, waving us over, shouting, laughing. It was the early 1990s, not a lot of mother-son kayakers out on the water. This was before "sea level rise" was a household word, another time in history as far as I'm concerned. One short wooden pirogue that looked like a sausage hollowed out was tied to a pier. They must have had a fiberglass *panga,* an open, outboard-rigged fishing boat for longer range, but it was out working.

The doorless shack was made of what looked like anything that might have drifted by, salvage wood and random pieces of sheet metal nailed and tied together. A young man, skinny in his ratty T-shirt, came out from behind an open window, ducking through a low entry, and from the water below we asked him for directions and the next good drinking water. He did not need to see our maps or compasses. He could have closed his eyes and told us where the tide was at any moment. Low waves glided up the piers and back down as we rose a few inches closer to the man and then fell a few inches away.

I showed him my map anyway, pointing out the island we wanted to reach next, and he nodded his head with authority, *Over there.* No cardinal directions or degrees, he knew exactly where everything was. When I peered to where he was gesturing, it all looked the same. When we pushed off and the hut disappeared behind us, again came that feeling I'd been cut free in space.

Later that afternoon I was relieved to see mangroves on the horizon, just as the young man had said. The first island was pretty much unapproachable, also as he had said. It was overgrown with mangroves, one side bayed into a shallow lagoon, a bit of damp sand where we could fix a meal. By the time we finished eating, the sun was dropping, and the tide was coming in, meaning that soon all that would remain here would be mangroves in knee-deep seawater. My mom paddled off for a slightly higher and drier spot within sight of our lagoon, a clearing just big enough for her to put up a tent. I stayed and strung a hammock over the water. For hours that evening I lay facing up listening to the sea's autoerotic lapping, my kayak bumping forward and back beneath me. I was expecting the tide to start dropping soon, but it kept rising. It climbed until sets of gentle waves began brushing the kayak against me.

I closed my eyes, listening to waves sluicing through mangrove. Why would a tide continue rising past its expected turnaround? Was there something I didn't know about the moon, or maybe a storm in the Atlantic beyond the barrier reef? Was the young man in his landless shack also worried, noticing a change he'd never seen before? (You think of fishermen in Southeast Asia jumping in their boats and heading for open water as the sea draws out just before a tsunami.)

In 2010, the Marshall Islands in the South Pacific were inundated by a tide that rose a full five and a half feet higher than projections of worst-case scenarios for sea level rises in the next hundred years. It was an epic event, brought on because of a warm La Niña cycle. Warm-water tides are generally higher than cold, and the position of the moon coincided with a thermal high to swamp parts of the Marshall's capital city of Majuro. When water receded, people put their lives back together, but a new benchmark had been set. Sea levels had momentarily set a record.

It does not happen all at once. It comes in stages, seas rising notch by notch. I switched on my headlamp and leaned out to check my markers, wondering if I was going to have to hop in my kayak, or if my mom was still dry. Driftwood and a green coconut floated in and out looking a little like a house flooded, furniture tipped over, books floating past. I checked my last marker, and it was above water on a higher swell. That's all I needed to see. The tide was turning, I just hadn't timed it correctly, ignorant landlubber. I rolled back into my hammock and fell asleep.

I swung back around to the coast on a gravel road heading toward the village. The north shore was taking a direct hit, rollers crashing over each other on this shallow coast. Locals told me this was nothing. I should be here for forty-foot seas, especially in winter when the village is being swept by flying ice.

I caught up with my mom out by the smoldering village dump, its gray smoke tilted nearly flat across the tundra. She was being pelted with sea mist and still looked cold, a new drop of snot dangling at the tip of her nose. Ahead was the cluster of Savoonga, the water edged by telephone poles and colored roofs about fifteen minutes away.

"It's July!" my mom shouted at ravens passing overhead, who looked down at her with quizzical authority. "It's July, get it? July."

That means, *What happened to summer?*

Locals on St. Lawrence Island had complained to us that everything seems off. Sea ice breaks up too early in the spring, both bad for hunting and bad for shorelines and villages usually protected the high waves of ravaging winter storms. They said the kinds of fish they harvested had changed in startling ways, species they'd never seen before coming up in nets and on hooks. The same is true for concentrations of seals and seabirds, as if an ecological switch were being thrown. You could say that even out here at the end of the world, Yup'ik natives have been oversaturated by sensational media, or they have become hypervigilant because of the increasing number of scientists landing to take measurements. But when they spoke, it sounded like just the facts. No haunting warnings, no headlines, no juicy quotes. They said that changes were coming fast.

Ahead of us we saw a collection of retired umiaks, walrus-hide boats up on shoulder-tall blocks. They'd been replaced by aluminum skiffs and outboards, better chance of survival while hunting in rough seas. Cinched down by ropes, peeled by wind, the umiaks pointed straight at the sea, and we ducked under one, getting out of the wind for a moment inside a curved wood-frame hull, salt spray rasping against its thick, leathery hide. Light came in through holes and thin patches, no longer seaworthy. The umiaks were becoming archaeology.

Between the retired umiaks and the sea lay decades of whale carcasses. Some were reduced to massive gray-white bones, while fresher harvests hung on to old meat where not everything could be taken, some left for ravens. Rancid blubber and head meat washed back and forth in the waves downshore from a jawbone feathered in golden baleen. These whales had been hunted by hand, tagged with harpoon, killed with rifle shots from a skiff. What bravery, I imagine, both men and women taking on a seventy-five-ton animal by hand. From under the umiak, I could see a gun-metal fluke sticking out of the sand tied off with frayed yellow nylon rope from where it had been towed into the bay. If you went out into the water, you'd find more centuries of whale bones beneath the waves, a history of life on an Arctic island extending across what was once beach, old villages submerged, former Savoongas forgotten by centuries, whale hunts thousands of years in the distance.

John Church, an Australian scientist with the Intergovernmental Panel on Climate Change, has synthesized multiple layers of global data and in a broadly accepted conclusion says not only is sea level rise currently a given, it is picking up speed. Satellite altimeters employed between 1993 and 2006 detected a mean sea level rise nearly twice that seen during the twentieth century. In turn, the average twentieth-century rise was one order of magnitude larger than that of the previous two thousand years. Even if measured in quarter inches annually, a doubling and doubling again says something about the future. If the current measured rise of eustatic sea levels continues, the rate works out to nine to fifteen inches per century, some parts of the world sloshing up to an expected two-foot rise, which would dramatically change the way high tides and surges flood along places like the Eastern Seaboard and Bangladesh. Savoonga, too, will go under.

Gravel streets turned this way and that through the village, old women puttering on four-wheelers, teenagers whipping corners like hot rods. Box houses lined up perfectly, sealskins and flippers out curing, electric lines and radio antennas howling in the wind. All manner of scrap lumber floated in pools of melted permafrost, and I saw my mother eyeing each find, imagining furniture she could build. Throughout the village, seal carcasses hung from lines. Baleen from whales was mashed into the gravel under our feet. Downstairs at the city building, a wood-paneled slapdash job, there was a loud game of late-night bingo. The cramped foyer where you are supposed to stomp off ice in the winter was crowded with people drinking coffee from Styrofoam cups, all of them Yup'iks. One of the men pocketed his cigarette the second he saw my mom. She had given him a hard time a couple days ago, telling him smoking is bad for you, from which I turned away in utter embarrassment. A couple hands came out, shaking mine. Welcome to Savoonga, they said.

The big, younger hunters standing in the door did not even nod at me. I came from the outside world. I was from a realm that generates cancers, carnage, and rising seas. PCBs left on St. Lawrence Island by the U.S. military during Cold War operations have yet to see any kind of substantial cleanup, while cancer rates in the island's two villages are through the roof, people with PCB levels in their blood six to nine times higher than people in the Lower 48. Cancer rates are not merely from the ground but from trophic toxins that come concentrated in the meat of walruses, seals, whales, and salmon they pull from increasingly polluted oceans. Humans, the top of this marine food chain, are some of the most poisoned meat in the world. Sea level rise is only one piece to these people's puzzle.

Some villages are seeking someone or something to blame, trying to find the faceless culprit that is rocking their world. Like Shishmaref, the Inuit village of Kivalina along Alaska's north shore is disappearing into the sea. In 2008, a lawsuit was brought by the people of the village against two dozen energy companies, the likes of Exxon Mobil and Chevron. The claim is that these companies are the largest contributors to global warming by way of emissions resulting from the fuel they sell. Kivalina contends that if somebody is going to pay for the village to be moved, it should be those companies. The people have been seeking $400 million in relocation costs as their shoreline is stripped away year by year, foundations perpetually undermined. In reply, the U.S. district court noted that their predicament stems from too long a chain of events to name single sources, saying, "In a global-warming scenario, emitted greenhouse gases combine with other gases in the atmosphere which *in turn* results in the planet retaining heat, which *in turn* causes the ice caps to melt and the oceans to rise, which *in turn* causes the Arctic sea ice to melt, which *in turn* allegedly renders Kivalina vulnerable to erosion and deterioration resulting from winter storms." Based on such a remote, if not direct, line of logic, the suit failed and is currently in appeal.

What the court failed to appreciate is that these many "*in turns*" are less degrees of separation than they are points of amplification. In the Arctic, you can see the effects of positive feedback. Warm temperatures cause a cascade where a degree or two means islands disappear and ecology changes. This is a planet of trophic dynamics; energy transfers from one system to another.

Regardless of who or what might be to blame, the clearest pragmatic option for endangered sea villages is to relocate to higher ground, which for thousands of islands around the world means somewhere far away. For Arctic hunting cultures raised on harpoons, guns, and an abundance of seabirds, shifting to the mainland would be a sort of cultural death. There is plenty of land in Alaska to have, but for an island culture, mainland is their own extinction.

A broad-shouldered young man with hunter's hands stared at me, his eyes following me as I headed up the stairs with my mom. I could not apologize to him for whatever is going on, for the cancers, rising seas, and vanishing islands. My mom and I were living on this planet, too. Our room was upstairs. We climbed the worn wooden steps over the din of laughter and bingo. It was about ten thirty in the evening, sun low behind speedy clouds.

Stripping off wet gear, we hung jackets and pants over the backs of chairs. Narrow windows cut out of pressboard paneling looked across the village, practically a high-rise view from one of the only two-story buildings in town.

"Did you see that one guy, the way he was looking at me?" I asked. "I've seen him before. I don't think he likes me."

Putting on her nightgown, and a coat on top of that, my mom said, "Of course he likes you. You're just imagining things."

She sat at the table with bare feet on a linoleum floor. Dinner was ramen heated in the office microwave, then bed. She is a punctilious woman, a fresh shirt almost every day, an application of lotion to her face before bed. She made me look like a ragpicker. Within half an hour, she was asleep on her cot, and I was standing at one of the narrow windows watching the storm slowly subside, passing south toward Japan.

Midnight. Outside, the wind blew and clouds skated past. Sunset glowed under the storm, or was it sunrise? This far north, hours were very close together, hard to tell one from the next. Bingo had shut down for the night, all the four-wheelers gone home. While my mom slept that night, I strolled out of town past the new school and beyond the two tall wind turbines near the village's junked equipment—a pair of rust-pitted bulldozers retired side by side as if inseparable even in death. The tundra beyond there was shrill with chattering ground birds—plovers, wagtails, stints. Birds did not seem to sleep, taking advantage of the long, somewhat warm side of seasons. I followed the road leading to the village cemetery, where whale bones and white crosses stood over graves. Some were in the process of spilling into ponds of sunken permafrost, human bones falling out of decayed wooden boxes. Large portions of the cemetery appeared to be leaning, a hill falling over as if its meeting with the nearby sea were inevitable.

I walked down to a bowhead butchered decades ago, its skull big enough to sleep in. I took in the scope of the sea as it dove against a shoreline corniced with crusty snowbanks. I ran my hands along the desk-sized vertebrae, wondering if in its lifetime this creature had detected the changes. Perhaps its song had echoed back from the island differently from year to year. Passages it could never swim decades earlier were suddenly ample for travel, incremental rises in sea level allowing a way through where it could not have gone before.

In 2007, a bowhead whale was killed off the Alaskan coast, and found embedded in the blubber of its neck was the tip of a harpoon dating back to the 1880s. That whale was estimated to be 130 years old when it died, pushing the life span for the species to about 200 years, longest-lived mammal on earth. Given such longevity, the slow-moving changes of modern sea levels might be apparent to these whales. Two centuries ago is when the current upswing of sea level rise appears to have begun. I wondered if in their migrations they are aware of the movement of ice sliding into the sea, if they notice and remember new shapes of currents and topography that humans require instruments and charts to see.

There are events we can now see with our own eyes. I once sat at a bay on the west coast of Greenland where the sizable Jakobshavn glacier was unloading into the sea. I remained in one place for thirteen hours, counting fresh icebergs coming off the tongued edge of the ice sheet. I saw nearly a hundred icebergs that day drifting through the bay, some as small as two-car garages, others hundreds of feet tall, sailing in the distance like floating mountains. Each I imagined forming into a bubble of freshwater and disappearing into warmer, saline currents that come up from the south. Each formed at the tip of an invisible eyedropper, adding to the change twenty-four hours a day.

Fourth of July is a big holiday here. Savoonga knows how to do it up right: hot dogs and chips on paper plates with folding chairs set out for the elders. People greeted one another with a cheerful "Happy Fourth of July." It was the biggest outdoor party of the year, given extra significance considering Russia is within sight and any other piece of Alaska is not. For the Yup'ik at this point in history, they celebrate the United States because it is the country to which they happen to belong. The fire department rolled out its one and only fire engine, a red 1960s classic, siren whining. Even the weather cooperated, bringing clear skies, the village brightened by light as rich as any Dutch painting. Feedback screeched through the village on a portable PA as people came up one at a time either thanking all those they were grateful to or delivering names of raffle winners. Half the time I could not tell if they were speaking Yup'ik or English.

The oldest members of the village—in their eighties and nineties with prune-wrinkled faces—kept on their jackets and caps in temperatures warm even by my standards. Toddlers were dressed in the same plump cold-weather gear, little faces appearing through oval ruffs as they wobbled around with arms sticking out. Everybody else was in shorts or jeans and shirts.

Children and younger teens carried plates over to the elders. At sixty-four years old, my mom counted as an elder. A Yup'ik girl dutifully brought her a meal in outstretched hands, stopping in front of her with a presentation of a Wonder Bread bun with a boiled hot dog, a pile of Fritos, and a couple packaged cookies. My mom smiled and said, "No thank you." She had already had a hot dog, and she had eaten it out of her hand. The girl did not move, puzzled, as if my mother were a species she had never seen.

"Take it," I whispered.

"I can't eat another one," she whispered back as the girl remained motionless and baffled before her.

"Take it. I'll eat it."

To the girl's relief, my mom reached out for the plate, thanking her with a sweet smile. When the girl turned away, my mom held the plate for an awkward moment, looking around. I reached over and took it from her.

Savoonga used to celebrate Fourth of July on a gravel spit that extended north from the village, a good place for sprints and ball tosses. That spit was now submerged, hardly visible at even the lowest tides. Now the celebration happened in front of the two-story town hall, bunkered from the sea by bouldery ramparts bulldozed into place. Bike races launched in front of us, kids in shorts and T-shirts pumping to the gravel airstrip and back, man on the microphone calling, "Come on, let's go, let's go. There are no losers here today, everybody's a winner." The kids and their bikes were all mismatched. A girl with ribbons tied to her handlebars barely reached the pedals, so determined to win she grunted at every stroke and came in a winded third. There were also footraces, quarter mile, half mile, and then some big loop that took the young men out to the tundra and did not bring them back for hours. Runners were coming in at all times and from different directions, late returns red faced and sweaty, a bony-legged girl wearing headphones and a gang shirt, the Anglo pastor's wife panting in spandex, and an overweight middle-aged Yup'ik woman waving at all the applause coming to her as she pounded down the road.

For a moment, this did not seem like an island in a rising sea. Life here felt permanent. Although plagued with poverty, alcoholism, and drug use, Savoonga was a vestige of a way of living important to hold on to. If New York and Hong Kong should burn, civilization collapsing, I thought at least this island should be spared. We will again need people who know how to endure. When a Savoonga boy told me about a hunt, he could not help throwing his arm as he spoke, acting out the hurl of harpoon without even thinking. An older man explained to me that when harpooning an eighteen-hundred-pound walrus through a hole in the ice, you've got to sink your barb into the beast's muzzle when it comes up for air. You let it go on a rope and for hours haul back a little at a time until you have worn it out.

I told him that sounded downright badass. He nodded in complete agreement. Few people where I come from could imagine taking down a big animal with a sharp stick and a rope. Softening into desk jobs and general obesity, to forget the hand of survival is to become a different species.

Laughter led to talk, talk to an invitation to come inside. A woman whose freezer was stocked with slabs of walrus meat and salmon served us Kool-Aid at a kitchen table with aluminum legs. Her husband was a middle-aged fellow, balding and slim. I gave the couple a small watermelon and a bag of cherries I brought as gifts from home. I pulled out tea. The man laughed that I was like Santa Claus, my day pack bottomless. While my mom spoke to the woman off to the side, he told me that the edges of his island were disappearing. He checked off locations that had visibly reduced within the span of a generation, beaches and coves now underwater. He flipped through photographs he had taken, showing me a familiar landmark, a beautifully arched sea stack just off the island's north coast. It looks like a piece of alphabet standing out of the water, ironically spelling the word "hi." A big chunk of the *h* had recently collapsed into the waves. He did not capture the actual fall, but showed me befores and afters, saying the change is not coming, it is here. He did not speak with a weary, solemn tone like a doomsayer, nor with the animation of an evangelist. He had sort of a grin on his face, his voice sharply intelligent, his eyes watching my every reaction. His grin asked if I understood. He was simply telling the truth. This is what is happening here.

My mom could live with this temperature. In clear Fourth of July sunlight, she was practically dancing out past the village. We walked together toward the headlands east of Savoonga beneath bevies of flying seabirds, so many murres and crested auklets in flight I could not possibly venture a count. Seas were calm, hunting perfect for the birds. We craned our necks, gazing up at clouds of passing birds. Life was pouring from earth and sea. With its cold, shallow waters and nutrients stirred off the seafloor into columns of sunlight, this is one of the richest marine environments in the world. Puffins launched from black palisades, skimming over waves, while in the village a mile behind us the fire engine horn blasted, followed by a cheer. Across the tundra, the sound was unexpectedly crisp. Another runner must have staggered back.

At the top of the headland, my mom wandered off on her own, then sat on a boulder, face tilted toward the sun. I dropped into a shallow draw cut into the basalt. A thick bank of snow lay down inside. Its curved eave dripped and drizzled onto rock, spattering, filling pools, running off toward the sea. The bank was melted back enough that I could crouch under it behind a dripping curtain. Different from my mom, I gravitate naturally toward shady places on sunny days like this, even in the Arctic. Sun shone through each drip, and I inched closer, seeing a prismatic upside-down world, sea and sky reversed. Each teardrop shimmered for a moment and vibrated tenuously, then fell. This is how climate works, I thought. Forces push and pull, weather begins to switch back and forth, summer and winter turned upside down, and then the system jumps. The drip falls.

During Egypt's Alexandrian heyday, the fertile edges of the Nile River delta were lined with busy and interwoven cities that have since disappeared. Two were recently excavated, Herakleion and Eastern Canopus. These were not land excavations; the cities were found underwater a mile from shore. Excavations were done by scuba divers. Among sunken temple columns and bronze artifacts, a rock slab was found inscribed with tax edicts and the signature of a pharaoh who ruled in the third century B.C., someone who probably could not have imagined his own rich corner of civilization covered in waves, especially since it had been built six feet above the highest tide to start with. The cities were built on soft Nile delta sediments that over time gave way to the weight of stone architecture above. Cracks formed beneath buildings in Eastern Canopus and were intentionally filled with sand and debris, suggesting people knew what was coming and were trying to stem it. In Herakleion, excavators found statues and columns toppled and walls strewn atop human skeletons. People had not been able to get out in time. Among these remains were valuable artifacts buried beneath rubble, bronze pieces you would have grabbed if you had time. But people obviously didn't have time. This was a story of a swift and unexpected collapse. The sediment mixed around these ruins is swirled with confusing strata. Judging these swirls, scientists have called this calamity a flood. An old channel of the Nile probably burst open. The ground liquefied as water-saturated sediments rose rapidly and literally consumed parts of the city. After that, occupation continued in some parts that remained above water, but these were scant settlements, probably desperate; you can imagine the refugee camps made of stone, the dead and the dying dragged out of rubble as survivors clung to their rocks.

The natural processes that affected these two Egyptian cities have not stopped. Only the stakes are higher, at least in our modern estimation. Venice, Shanghai, Bangkok, and New Orleans are all built analogously on soft delta sediments. If lower parts of New Orleans had been simply abandoned after Hurricane Katrina in 2005, levees left unrepaired, their remains could have appeared as compelling as Herakleion and Eastern Canopus, human skeletons found in flood-buckled neighborhoods sinking inexorably into the Gulf of Mexico.

How do you fall in love with that which erases the world, either fast like Herakleion and Eastern Canopus or slow like the last several thousand years of Arctic islands pinching out, and on them fauna genetically adapting as fast as it can, until finally there is no more room and species perish?

Any change seemed especially hard to swallow on a sunny July day. Sometimes you just want to relax and let the world spin on its own, let the drips fall. I leaned back inside the chamber and rested my head on my day pack, closing my eyes to the loud sweep of birds outside. When I woke, I didn't know how long it had been. I climbed out of the shade and walked to the point of the headlands, looking straight down a cliff past colorful, nervous puffins looking back up at me. The sea below was turquoise and calm. Honestly, I did not know the Bering Sea ever settled down. Sun-rippled shoals appeared beneath the surface, stretching a hundred yards out before falling into the midnight blue of slightly deeper waters.

I looked for my mother, not sure where she might have gone. When I retraced to where I last saw her, she was not there, disappeared among gullies of black rock and squawking chicks. After half an hour, scanning distant tundra, I still didn't see her. I wondered if she might have fallen off the edge of this palisade, slipped on loose rock, and plummeted sixty feet to the water below. I would not put it past her. Instead, I found her nearly camouflaged and hunkered down between rocks with her hands on her knees, her face perfectly set in the sun with that same small smile.

She glanced at me, closed her eyes again, and in a dreamy voice said, "I love days like this."

"I wondered where you'd gone," I said.

She cracked an eye, looked at me that way she did only when examining my whole face. "You need sunscreen," she said.

After several days, the village council agreed my mom and I were harmless, giving us our permit to travel beyond the three-mile radius. Within the hour, we exchanged a handful of cash to borrow a four-wheeler. Grinning Man said he'd take us out with him on his hunting rounds for a small fee. It was better than the embarrassment of going alone and bogging down a quarter mile from the village. Gunning to keep up with him, I leaned over the steering frame while my mom clutched on to my back. The weather turned days ago, Fourth of July over. A storm filled the space with wind, rain, and low clouds streaming out of the north, the Arctic breathing down on the island once again. Grinning Man knew every twist and turn of tundra, which, like the sea itself, was surprisingly uneven. My mother's face nestled at my shoulder so she could see where we were going through streams of blowing mist. Grinning Man was so far ahead I could barely see him.

No road out here, it was nothing but a route, tires grabbing at slicks of bog grass. Whenever we hit a dip and I rammed the accelerator to get up the other side, my mom unleashed guttural, involuntary sounds. Fat tires bounced through holes, and she bellowed, "*Argh, ungh!*"

She jawed into my shoulder, groaning, "*Slow down, oh God, oh ghawk!*"

"Stop it!" I shouted at her.

"Well, slow down!" she shouted back.

"Can you just stop shouting in my ear?"

"I can't help it!"

Before us spread a green and rolling expanse, more or less what the land bridge would have looked like in its day, no trees and a lot of open ground. This is how people must have crossed this place; I can see it clearly fifteen thousand years ago. A man was moving across this ocean-less slope with his mother riding his back. Her bony finger pointed over his shoulder warning of every gully and mammoth ahead. Her voice rang in his ear, *Watch out, be careful!*

After an hour, I started to feel as if I knew the machine pretty well. I was also feeling wet seeps inside my coat and pants, my boots drenched. Grinning Man turned toward the sea and followed a cobbled shore at the high-tide line. As I came in behind him, my tires crunched and dipped over whale bones and driftwood, and my mother sounded as if she were being beaten. Grinning Man stopped at the skeleton of a bowhead rolled up from high waves. By the time we caught up and shut off the engine, he had already crawled down inside the skull. There were old ropes on the ground, a few rusted cans. The whale had been harpooned and brought to shore maybe twenty years earlier, butchered here, meat hauled back to Savoonga. I walked over as my mom dismounted in relief.

Grinning Man sounded as if he were working inside a car engine and came out of the skull with two small bony globes in his hands.

"Whale cochleas," he announced with the biggest, most toothy grin yet. "Ear bones. They sell for five hundred dollars apiece." There was no greed in his voice, just a good take for a man making a living by hunting and gathering. He must have had claim on this skeleton, waiting for the right year until scavengers and weather had cleaned it up.

While he went back to his machine and bungeed his finds to the frame, I ducked inside the skull. These are big-headed beasts, *Balaena mysticetus*. The head tends to take up a third the length of the entire body, bones grown massive to break up through sea ice to breathe. You could comfortably seat two inside its skull. Looking into the clean, wind-hollowed braincase, I wondered if I could trade my own decades for a two-hundred-year life span just to see what page turns next for the earth.

Grinning Man was calling for me. We had ground to cover. After my mom got back on, we looped out to the tundra around a headland, following windy, wet flats out to a narrow cape. Along a boulder shore he eventually stopped and unstrapped a shovel. He was digging an ancient village for artifacts and fossil ivory to sell. Already we had passed a number of old villages, abandoned rock piles so old they had turned into nothing but humps of tundra and old wall lines half in and half out of the water. This was a big one, and it was absolutely cratered, bones pulled up everywhere by a generation of diggers. He said that different families had their own plots that they work. The same way you would not steal bird eggs from someone else's rookery claim, you would not think of digging someone else's ground. His spot, where he had been probing since he was a kid, was along the wind-sprayed shore, old village walls rambling down into the waves. He climbed into a muddy hole and started throwing rocks over his shoulder.

Fossil ivory found buried in archaeological rubble makes fair income, a handful of good pieces covering a month's groceries at the village store. You look for whole walrus tusks left behind from butchering grounds two thousand years ago, permafrost and soil having impregnated the ivory with a rich ocher color, perfect for carving. The real prize, though, is a high-quality artifact, maybe an ancient pair of scrimshawed snow goggles made of whale bone and turned a golden coffee color from time in the ground. Caches of artifacts found along these coastlines have garnered $200,000 a load, although most of what you find is just splinters and little chunks. You can hardly take a walk in Savoonga without someone approaching you with a plastic bag full of busted artifacts to sell, all manner of broken harpoon heads and chips of carved ivory, ten dollars each.

People here say that these remains are gifts from their ancestors to help them live in a cash economy. I doubt if their ancestors could have imagined this cash economy with its Carhartts and four-wheelers, perfectly good umiaks put up to rot on blocks while hunters motored out to sea in aluminum skiffs with rifles and steel-tipped harpoons.

The ground was littered with hundreds of ancient, discarded walrus skulls. They had been dug up by modern artifact hunters and pushed aside. Whale ribs stuck out of the upheaval of dug-up tundra, and I walked among them, finding minor, discarded artifacts as numerous as pretty shells on a beach. There were bone-scraping tools, seal ribs beveled and drilled, and whale-rib sled runners that would have been lashed onto snow sleds with sinew. Grinning Man had told me the best finds are not on land but out there, in the water. The shallow coast is lumped with old, submerged villages. You take a boat and a compressor, put on a dry suit and a mask, dive down with a tube to breathe through. Sometimes you find the real old stuff out there, Okvik burial figures made out of ivory. Is any more proof needed that sea levels rise? This island is tectonically stable. It does not move up and down. It is the water that moves, the Bering Land Bridge exposed and covered many times over the last few million years.

I walked over hillocks of bones, fog racing out of the north, breath of the Arctic spinning across this shoreline. Mist thickened until I couldn't see our four-wheelers. Grinning Man was lost in his hole, and my mother had disappeared among the ruins. Every gray stone became a ghost. Then my mom appeared standing on the highest rock like a short lighthouse, her arms pulled in, walrus skulls strewn around her. Grinning Man had climbed out of his pit, not finding anything worth keeping. He walked back to his four-wheeler, where he took out a battered thermos of hot coffee and carried it to my mom, pouring a lid-full for her. She drank, passed it back, and pulled from her coat pocket a package of peanut-butter crackers to share. I had a greasy bag of nuts to add to the collection. Grinning Man produced Pop-Tarts, one foil package for each of us. Together, we ate and drank in threads of fog, warming ourselves in a muddy field of skulls.

4

Civilizations Fall

The world dies over and over again,
but the skeleton always gets up and walks.

—Henry Miller

PHOENIX, ARIZONA

THEY CHANGED the name of my mountain.

Smack in the middle of the city, this charcoal-colored peak was the pole around which my childhood twirled. Born and somewhat raised here, I have had it burned into my memory, a watchtower, a desert oracle. Its splinter-headed hulk of Precambrian desert rock rises a thousand feet above Phoenix.

Not that a simple name change was such a big thing, I just liked the name the mountain had, the one I grew up with, Squaw Peak. Now it was a mouthful of Hopi phonetics: Piestewa Peak, pronounced "pie-es-too-ah," the *t* almost a *th* but not quite.

I'm heading up Squa . . . er . . . Piest . . . er . . . wa Peak.

It was a summer evening, and I was hiking for the summit. The ground was warm as an oven from the day's sun just set. Laser lines of twilight faded on the western horizon over the slow burn of the city. I've always come up this mountain, a refuge, mostly at night. Days can get crowded, a hundred thousand hikers a year going for this horny summit. At night, you could see forever, a tapestry of streaming freeways and vaporous streetlights spread all around, growing more expansive the higher I climbed. It was a city with no end that I could see, *Los Angeles, Tucson, here we come.*

The city used to be smaller, half this size when I was a kid, its airport a Podunk once bearing a quaint and silvery thumb of a

115

control tower. Tonight planes were stacking up in the sky, seventy passenger-jet landings per hour.

I used to scramble up these inner-city mountains in high school on Saturday nights, coming out with friends to dodge police helicopters. The Phoenix Mountain Preserve closed after 10:00 p.m., and around midnight we'd huddle and hide as spotlights blazed across statuesque saguaros and boulders pecked with ancient spirals, lighting them for an instant, then plunging them back into darkness. I remember standing on top over a hilarious ocean of light, feeling like a young squire surveying his future, a land full of allure and promise. Tonight the sky glowed as if under its own power, all but the brightest stars completely blotted out. It felt brooding, dangerous even. I came tonight to make peace with this new name.

"Squaw" means "Indian bitch." It means other things, too, all about women and their privates. You can see why you might not want a name like that standing at the proud center of your city. *And here, my fair friends, is Indian Bitch Peak.* Or Squaw Tit if you'd rather, as it was called by early Anglo residents around 1910. Considering the dreadful killing and treatment of Native Americans in the first two and a half centuries of this nation's existence, the name was just callous at this point. The Hopi name, Piestewa, is a blessing word that calls water to this place. You can't beat that in a desert city.

Dry summer thunderstorms rimmed the broad desert basin of Phoenix, lightning dashing through a perimeter of boiling clouds, but there was no rain. The air was parched. It smelled of rock and the simmering rise of burning fossil fuels. The last of the day's hikers several hundred feet below were just getting down, their flashlights and headlamps winding toward the parking lot. The dull groan of the city was broken only by peals of motorcycles or barking dogs. Every once in a while I'd hear a gunshot, maybe gang related, more likely some yahoo in his backyard. My dad used to do that when I lived with him at the north end of the city. He'd point his .357 straight up and unleash a sudden flame into the night, as if trying to blow a hole through the sky, an exit to crawl out of. There was no telling where that bullet would land.

The trail fishtailed up through crooked hunks of rock. You could see just fine without a flashlight, everything bathed in a faint lemon-colored glow. As you wound up the ziggurat of the summit, views opened to other sides of the city. It had no center, no blazing core like the Pudong district of Shanghai or Manhattan in New York. Instead, Phoenix was a carpet of ember-bright lights, cords of traffic, and the small, brilliant clusters of ballparks. Freeways moved like slow arteries. There was nothing fast or particularly glorious about this city, just inexorable.

Not a great city by any means, average in size at four million people, it is located in one of those settling places, a region where people have always come. A 2006 excavation for a new convention center in downtown Phoenix exposed a village of forty pit houses dating back three thousand years. Two intermittent rivers converge from out of the highlands in the east, flowing across a broad subtropical desert, allowing for a nearly year-round growing season. Look down at this place at night, and you can see the blueprint of its evolution. The greater metropolitan area has not a central downtown like most modern cities but a web of satellites, the cores of Tempe, Mesa, Peoria, Scottsdale, and Phoenix proper. This is a hydraulic blueprint, an original pattern of irrigation canals that set the layout of the city.

The phoenix is a mythical bird, Asian and Middle Eastern in origin, from a fantastic tale of death and self-generated rebirth. This legendary flame-colored bird is said to ultimately rise up incandescent as its nest ignites. Fire consumes the creature, and from its own ashes a new bird is born.

The name is for a reason. Lift the sprawling rug of this city, and you will find an ancient city buried below. There were earthen temples, roofed villages, plastered ball courts, and a vast network of irrigation canals and ditches, all predating modern Phoenix by at least a thousand years. People of Native lineage grew cotton, corn, beans, and agave here long before there was ever a Columbus to set sail. There were farmers, hunters, traders, feathered priests, and fine-handed shell carvers exchanging their goods hundreds of miles away. Equal in complexity to early Mesopotamian society, this was the beginning of a civilization, a first shot at grandeur that has taken root at one time or another on nearly every inhabitable place on earth. The people who lived here are now called Hohokam (accent on the first syllable). The word is a bastardization of the O'odham word *huhugam,* meaning, simply, "ancestors," or "ones who have gone."

When Anglo settlers started setting up shop in the Phoenix basin around 1860, they found ruins and mysterious fields of broken pottery left from Hohokam ancestry. Every new canal that was built fell into the path of these much older ones already buried in the ground, same size, gradient, and direction. It is as if we poured in a new civilization: Just add water, and the hydrologic ghost of the Hohokam comes to life. Between here and Tucson there still stands a thousand-year-old mud great house, a three-story ruin like a rook on the chessboard of the desert, modern neighborhoods, prisons, and convenience stores crowding in around it. Deep inside the city, eroded walls stand, marking chambers erected atop an earthen platform, the sun still rising through its rooms each morning. The Hohokam had bloomed for many centuries, but in the end ruins like these were all that were left. Upon them, nineteenth-century settlers called their new city Phoenix.

In the lights below, I looked for the old Hohokam blueprint, seeing clusters of strip malls and office buildings around what were once headgate communities. The dried-up course of the Salt River was a stream of light, a memory of water. In the distance, I saw the outliers of Queen Valley and Goodyear, places the Hohokam also occupied. Modern excavators have dug up pots and skeletons in these places, layers of boom and bust, bones pocked and weakened by diseases, malnutrition, violence.

This is a version of the world we greet with both fear and fascination: the collapse of a civilization. We are captivated by it, because no matter how confident we are in our modern infrastructures, we know it is possible. It is one of the options on the table.

I wonder, did historic settlers know what they were doing when they gave this city its name? Did they realize the narrative of rise and fall they were entering themselves—and us—into? Death and rebirth, and by fire no less.

This is our track record: cities left like empty shells around the world—Palmyra, Machu Picchu, Teotihuacán. Pre-Columbian temples and intricately carved slabs of stone push up from under the streets of Mexico City, while London coughs up Roman skeletons and two-thousand-year-old bronze tableware. We live in the graveyards of former societies. This is the song we've been singing for thousands of years, call and repeat, glory and decline.

This time, we say we have it figured out. We will not fall. As if we haven't heard that before.

Helicopters patrolled the distance, making the city feel like an enormous prison, spotlights hounding through backyards and alleys. The higher I climbed, the more I could see. There was still a halo of darkness out there, places at the edges not yet filled, though lights were breaching south toward Tucson and west toward Los Angeles. What once were interconnected lakes of lights as I was growing up were now one scintillating ocean.

On a butte much smaller than Piestewa, farther east into Tempe, I once sat with a friend, a contract archaeologist named Tom Wright, a digger who had worked in and around the city for thirty years. His job was to remove ancient objects in the way of development, pulling up pots and stone hoes, either returning them to tribes descended from the Hohokam, putting them into storage, or feeding them to museums. The round-topped summit we'd hiked to was decorated in ancient rock art and surrounded by sky-dazzling stadiums and freeways.

As dark came on, Wright and I talked about how civilizations tend to fall, common themes you see throughout time: environmental decay, failure of top-heavy infrastructure, resource depletion, loss of social egalitarianism, disease, conflict. Angkor Wat in Cambodia experienced hydraulic failure in the twelfth century, its irrigation network and retaining ponds proving too vast to maintain, collapsing just as the region became an agricultural hub for over one million people. After that, the area was left in ruins. Ur in Iraq lost its port during a drought in the fourth century B.C. when the Euphrates River shifted, leaving the city far from the water's edge. Every age, region, and society has its unique but similar cause and effect. Lop Nur dropped from Taklimakan trade routes and was abandoned when sea routes bypassed the desert. Memphis in Egypt declined against the backdrop of new power centers as its temples were disassembled, the stones carried away and used to erect new sites.

As Wright numbered the usual suspects among failed complex societies, I pressed him, asking what Phoenix would actually look like if the balance tipped and everything we knew slid into oblivion. Personally, I was thinking scattered fires flickering across this broad valley and dogs barking in the distance, none of this roar and glimmer.

Wright answered, "It would look like this."

It wasn't the answer I'd expected. But when he said it, it made perfect sense. He was saying you probably wouldn't even know.

Little by little a city falls, foreclosures eating away at the edges until you have eighty-eight thousand houses standing vacant.

Wright and I were looking down upon this blossoming city just as an economic plunge had been set in motion. Land that had been ninety dollars a square foot dropped to nine dollars within months. Construction on many developments would come to a standstill, neighborhoods left half-empty, partly constructed, as moratoriums were put on building new schools. Swimming pools on foreclosed properties turned into trash dumps. Some people left in a hurry, their money gone all at once, medicine cabinets still stocked with toothpaste and aspirin, dishes in the sink, take-out boxes strewn on the dining room table.

The postapocalyptic scenario of people living in the fresh scrap of their own collapsed societies is more Hollywood than reality. Collapses tend to take time, an evolution in themselves. Rome wouldn't have known it was falling even as poverty and decay spread, aqueducts failed, schools dissolved, and armies stretched untenable distances. The city was still standing, still occupied, the whole while.

Depending on which historian you hear it from, the fall of Rome took anywhere from three hundred to seven hundred years, which equates to more time than, say, the United States has been in existence. Slow it down enough, and you'd never even notice. In his classic eighteenth-century work, *The History of the Decline and Fall of the Roman Empire*, Edward Gibbon wrote that the Roman Empire's fate "was the natural and inevitable effect of immoderate greatness. Prosperity ripened the principle of decay; the cause of destruction multiplied with the extent of conquest; and as soon as time or accident had removed the artificial supports, the stupendous fabric yielded to the pressure of its own weight."

Two hundred and fifty years after Gibbon wrote his opus on Rome, Thomas Homer-Dixon, political theorist and futurist at the Balsillie School of International Affairs in Waterloo, Canada, wrote a corollary for our own scale of civilization: "If we try to keep things largely the way they are, our societies will become progressively more complex and rigid and, in turn, progressively less creative and able to cope with sudden crises and shocks. Their collapse—when it eventually does happen—could then be so destructive that there would be little of the prior order left behind. And there would be little left to seed the vital process of renewal that should follow."

Heading up Piestewa, I stepped through rock dust, my fingertips touching boulders and rough fins of dense, scaly rock until I reached the summit. A cooler breeze blew up here, worrying an American flag someone had jammed in a rock crack.

It was not the kind of summit where you could hold a dance. It was a broken pike sticking into the sky, no comfortable place to sit. The city washed up all around me, darkened only by the other small islands of inner-city mountains.

"Piestewa," I said out loud, like practicing in the mirror.

On the *t* my tongue nearly touched the back of my teeth, and I drew out the *wa* as in "want," as in "water."

The lineage of the name Piestewa traces back to old societies, water songs from long before Anglo settlers ever arrived. Though the name is Hopi, by definition you would also call the name Hohokam, "ancestor." When Piestewa officially became the name of my mountain, and I knew I'd have to say it, I contacted an acquaintance on the Hopi reservation in northern Arizona, inquiring if he would ask around and find a definition of the word. He returned saying that Piestewa roughly translated to "sitting at the happening place of water." It relates specifically to a flute society in which dancers would travel to a sacred spring where they would chant prayer-songs and play flutes to the water. Then they would return to the village plaza, where it is said they brought rain.

The name of Piestewa comes from Lori Piestewa, a Hopi woman killed during combat at the onset of the Iraq War in 2003. She was also the first woman from U.S. infantry to die on the front line of that particular war, a noble but rather unfortunate distinction. Piestewa had been a northern Arizona girl, mother of two from Tuba City. She worked as a Humvee driver for the 507th Army Maintenance Company in the first wave of the invasion. According to the official report, her convoy became separated and then lost in the desert and was ambushed in Nasiriya, where she was hit by what army investigators called "a torrent of fire." Piestewa was mortally wounded while jamming her foot down on the accelerator to get her and her company out of there. Her body was later found buried in a shallow grave behind an Iraqi hospital.

I whispered, "Piestewa."

The word did not sound like the carbon mass of a mountain that I knew, but it did have a ring to it, and I said it again, a little louder.

Piestewa.

It meant "Prayer for rain"; "Mother of two found buried in the desert"; "Warrior sent far away."

This is one way a civilization survives the longer ravages of time. It remembers itself beyond what is on the surface. It carries its own history ahead. I climbed down off the summit, monkeying across ridges into steep, loose ravines, letting her name fall from my mouth, memorizing the length of its vowels, a slow chant over this city so lovely tonight.

·   ·   ·

The first city to hit a million people was Alexandria around 100 B.C., followed in A.D. 700 by Chang'an, and then Baghdad by A.D. 1000. Each of these cities also experienced marked declines and total, regime-burning changes at different times. Ancient Alexandria was frequently sacked by invading armies who set the voluminous library ablaze, and it is now the second-largest city in Egypt at just over four million. Chang'an, which stood as the capital for eleven Chinese dynasties, some of which went down in fire, is now Xi'an, a mid-sized Chinese tech city of about nine million where renowned terra-cotta soldiers stand guard over the memory of their second-century emperor. Baghdad, meanwhile, was early on seized by Mongols and Turks for its strategic position, later declined under the Ottoman Empire until it was captured by the British in World War I, and fell once more to U.S.-led forces in 2003, its history nowhere near over. It fact, it may be only beginning.

I have a peculiar way of looking at time. It comes in all shapes and sizes. I know places where you can find pieces of broken Hohokam pottery a thousand years old—empty weedy lots by the airport and on the boulder flanks of certain buttes. You pick up a potsherd from the ground, dust off its face with your thumb, and peer right through history, centuries made of air.

Under the final approach to Sky Harbor International Airport in Phoenix, I walked around the excavated graves of mothers and children, a salvage archaeology project clearing the way for a new light-rail station. When you see women and children all buried together, it is a bad sign. I passed a hole in the dry, buff-colored soil where the bones of an infant had been laid upon the chest of a woman, the Hohokam mother, I imagined, the two of them placed in the same grave as if to keep them together forever. Another grave had held the remains of a young woman with an S-shaped spine, a bone disease related to malnutrition that would have left her crooked for life. After seven hundred years of strong occupation in the broad valley of Phoenix, the Hohokam died out.

This is what the end looks like, the final gasp of a civilization. Beneath the dropped landing gear of passenger jets screaming overhead, archaeologists had been dispatched to accumulate the dead with trowels, dustpans, and paper bags. The construction workers who preceded them had thought this was going to be a simple job, removing an old parking lot with jackhammers and backhoes to make way for the new development. Instead, they uncovered the tragic end of an ancient settlement. I came to a depression the size of a kid's body. I could not help crouching at its edge. The skeleton had been removed for transport and reburial, the hole left in the ground just big enough for the body—a young girl, I learned, early teens—her legs long and narrow. The excavation notes described shell bangles and bracelets on one arm. The remains of someone beloved. Almost two hundred bodies were found beneath that parking lot, marking the demise of the Hohokam in Phoenix around A.D. 1300, when the once busy heart of this valley was left eerily vacant for centuries to come. The Hohokam population had risen steadily through centuries of more or less excellent growing conditions. Its climax, numbering in the thousands, coincided with a debilitating drought. Violence brewed in the hinterlands as social networks and trade routes unraveled. Cliff dwellings near the headwaters were besieged, pueblos in more open country sacked. As if taking shelter behind their wisely built infrastructure, Hohokam settlements contracted, people crowding into smaller spaces. Irrigation demands increased as the soil became rapidly more saline, less fertile. Hunting became difficult, the game animals smaller and more scarce by the decade. Infant mortality skyrocketed. Drought mixed with overpopulation. By A.D. 1450, this was no longer a place to live. Those few who somehow survived walked away, their diaspora carrying off the last memories of this place, whatever happened here tucked into native legend like a warning.

The late Claire Russell and her husband, W. M. S. Russell, two distinguished British academics, studied the history of crashes in human civilizations and then compared them to what happens when other animals are knocked back by their own overpopulation. Not surprisingly, we don't stray much from natural cycles. "When a mammalian population becomes dangerously dense," the Russells wrote, "there is a reversal of behaviour. Co-operation and parental behaviour are replaced by competition, dominance and aggressive violence, leading to high mortality, especially of females and young, and a reduced population. The stress of overpopulation and the resulting violence impairs both the immune and the reproductive systems. Hence epidemics complete the crash of the population, and reproduction is slowed for three or four generations, giving the resources ample time to recover."

The height of the Hohokam civilization represented ancient America at one of its finest hours, with sophisticated hydraulic infrastructure and more than a hundred public ball courts within what are now the limits of greater Phoenix. But the Hohokam experienced a crippling end, a time of starvation, their bones showing signs of diseases that come from malnutrition. The final blow was a series of floods that roared down the drought-stricken Salt River. The last functioning irrigation systems were destroyed, the headgates gone. Canals and ditches were blown out and packed with silt. Too few people remained with too little expertise to make it happen again.

This is how it appears to have finally ended for the Hohokam: with a bang and then a fading whimper, the cry of a last infant in a village standing empty, its name never to be remembered again.

We have invented something new out of ourselves since the Hohokam. Beneath our pedestrian lives are grids of water mains, electrical conduits, and Internet technology cables.

Click-clank, the tunnel echoed.

I looked up. A twelve-foot passage opened above me where a bit of light leaked in through the seam of a manhole cover. A car had just driven over it. My headlamp shone around the circle of black widow webs. I was inside the city's waste system. You could walk for miles down here, Phoenix and Tucson nearly touching underground, anticipated to be a mini-megacity by 2050. This sewer was built after the 1980s, few cracks, pretty good condition for invisible infrastructure. When I looked ahead, my light faded down a tunnel so long and straight it tricked my eye, appearing to curve away slightly. I was in a wormhole, a dry tube six feet in diameter and straight as a ruler. What street was I beneath? Dobson? Oracle?

I was not alone down here. God no, I had a friend along, a Native arts dealer out of Tucson, a man named Keith who used to run the old Silverbell Trading Post. Keith and I came down here together every now and then, a pair of fortysomething urban spelunkers not yet free from high school curiosities. Our goal tonight was to find a new piece of graffiti that had been spotted, a brazen Hopi-like mask back inside one of the big junction boxes. So far tonight, we had just found the usual garbage made by potheads with spray paint and places where devil worshippers had been dripping black candle wax, leaving matches and empty lighters. Floods are pretty rare, so it can take a season or two to really give these waste systems a good washing. Still, they were fairly clean, only the occasional little chub of human feces left here or there.

I hadn't seen Keith since I started down this passage about fifteen minutes ago. He had taken another tube, but you can't get too far apart as long as you remember which way you are going. I shut off my headlamp, peered back for his light. Nothing. No sense of up or down, or anything but space. I stood for a moment in this black and starless passage. With no sound of cars, I figured I was under a housing subdivision, somebody's kitchen probably straight overhead. I flicked the light on. I reoriented myself and kept going.

If overpopulation, resource depletion, and climate change conspire to unravel civilizations, it is infrastructure that holds them together. The issue here is upkeep. You don't worry about what you don't see. The East Coast water infrastructure, put in by expert craftsmen more than a century ago, is in a state of system-wide deterioration. New York City's water main system is feared to be near collapse, with large-scale underground projects feeding more and more water into a rickety delivery system, some of the primary conduits still made of wood.

This is how a city falls apart, piece by piece. Cast-iron pipes from the late nineteenth century—employed in most eastern cities—have an average life span of about 120 years. Cheaper cast iron used in the 1920s lasts only 100 years, and pipes installed after mid-century are expected to function for 75 years. Put all those numbers together, and you are looking at, well, now.

About 240,000 water mains burst every year in cities and towns across the United States, mostly from age. Meanwhile, engineers have deemed thirty-five hundred U.S. dams unsafe, while half the country's navigational river locks have been rated "functionally obsolete." Most bridges in the country were built in the 1960s and 1970s, and they were designed to last fifty years. You do the math. A transit bridge in Minneapolis failed under ordinary load conditions during rush hour in the summer of 2007, killing 13 and injuring 145. It was forty years old.

But how fast or slow does this unpinning of public infrastructure happen? Based on the Hohokam analogue, the failures accelerate the closer you get to the end. Had those Phoenix floods come earlier, during the height of this burgeoning civilization, the damage would have been repaired, the losses cushioned, and life might have gone on. But in the fourteenth and fifteenth centuries that did not happen.

One pressing question is, how many people can there be this time around? What is the carrying capacity of our current civilization, of the land, air, and water on which it depends? We are not ancient Hohokam, numbering in the thousands in this valley rather than the millions. We are not even the Americas anymore, or Asias, or Africas. We are, as the song goes, the world.

Can we, the current multiethnic, poly-lingual civilization of earth, even compare our trajectory to the rises and falls of earlier people? Everything about us looks different, from our buildings to our clothes to the tools we carry. We sit at outdoor cafés in Copenhagen, eye level with the sea, and gaze through high-rise windows across the gray smog banks and concrete-colored buildings of Chengdu. Our jetliners crisscross the Arctic Circle, connecting Paris to Salt Lake City, carrying an ever-growing global consumer class. I would argue that we can be clearly compared. The fundamentals are still the same. We still put food in our mouths and breathe this very air.

We rely on the physical constructs of our civilization to deliver supplies and to keep grocery shelves stocked and gas stations pumping. The stakes have certainly changed, the ante upped by the billions. Early populations manipulated landscapes on a much smaller scale (the first atmospheric signs of civilization have been recovered from Greenland ice cores, where layers of iron soot correspond to smelting in Carthage from 600 B.C.). Now we are sucking up the slurry of Alberta's tar sands for oil and decapitating Virginia mountains for coal, damming nearly every last river for hydropower or water storage while rapidly altering atmospheric chemistry to levels not seen in millions of years.

If the question is carrying capacity, this is the moment to raise it. Within this half century we are projected to increase from 7 billion to 9 billion. Within a hundred years, most estimates put us between 10 and 15 billion, and the UN projects that Africa alone will go from 1 billion in 2010 to 3.6 billion in 2100. Based on population trends, the need for food goes nowhere but up. Within a thousand years, our numbers could be in the hundreds of billions if the population curve follows a perfect and unerring rise. Not possible, simply not possible. Sometime long before that, there will be a change.

There has to be.

Right?

No single curve holds its trend for long.

I slid my hands along the smoothed enclosure, until my light widened into a junction box and I stepped into a concrete vault some ten feet tall. Streams of chestnut-colored cockroaches fled up the walls, spreading to the ceiling over my head. What looked like a suicide note was written on the wall with a fat ink pen, and next to it a sloppy pentagram, or was it an anarchy symbol? On the other wall, flecked away by wastewater floods, were the following words in black spray paint:

Feel thy name annihilation desolating.
Hail of fire.

Metallica lyrics. You see stuff like this all the time down here, a bunch of apocalyptophiles we are.

I waved my light over the ceiling, shooing cockroaches out of the way, to see the silvery gasp of a crazy face painted in the corner, crude, certainly not gallery work. The mask wasn't in this room. I was starting to wonder if we had bad directions. It was Keith's son who actually found it, a young man who was in high school. He took a picture of it and showed it to us, knowing we'd want to go in to see it for ourselves.

I looked into the mouths of three other tunnels, wondering if Keith was in one of them or if an unexpected elbow in the system had taken him in some other direction. No sense waiting here for him, though. We had a mask to find. I peered into the two round, head-tall entryways going in the direction I was moving. In one, my light caught a pocket mouse bounding the other way. I followed.

It was just tall enough so I could move without ducking, ceiling curving about two inches above my scalp. Some of the passages I'd followed in the past were nerve-racking, small feeder tubes where you have to walk like a crab. This felt relatively spacious, though still claustrophobic.

My footsteps echoed as I moved at a healthy pace, concentric rings of my light dissolving at infinity. I tried to keep my imagination clear of any flood visions down here. It was summer, after all, flash flood season. Monsoon clouds had been gathering at the edges of the city, but we checked the skies before coming in. An hour in, an hour out. Knowing weather in these desert basins, we seemed safe.

But my mind wandered, and I couldn't help thinking of the Hohokam and the floods that did them in. A prayer for rains at the wrong moment could be disastrous. I saw this tube choked with shopping carts and old couches, half clogged to the top with impounded sediment, no one here to attend to repairs. That's the Hohokam model, at least.

Are we different? Our infrastructure is larger. Charlie Ester, the lead hydrologist for the Salt River Project (SRP), a public utility company supplying half the city's water, sat me in his office and showed me charts: water demand increasing, water availability decreasing. It was his job to make sure the right amount of water arrived, calculating years in advance, balancing reservoirs, opening and closing dams chained together through mountains and canyons. "The Arizona Department of Water Resources recognizes the SRP as providing a hundred-year water supply for new developments within its service area," he said. Then he told me he didn't think it was possible. He could not make such a promise. Some years, the water supply was not as robust as people thought.

I had flown with Ester over a quarter of Arizona, surveying watershed conditions east of Phoenix, part of the extended infrastructure keeping the city alive. Though he received data from remote gauging stations, he preferred coming out in person to eyeball the water situation.

Looking down through the helicopter's Plexiglas bubble, we followed rivers and creeks that all ultimately led to Phoenix. He signaled the pilot to take us down the crook of a bedrock canyon so he could peek inside for a sparkle of water. None, none, none. It was 2009, thirteen years into a stretch of a persistent and rarely interrupted drought, a period of water deficit rather than surplus.

We skimmed over Roosevelt Reservoir, the biggest of his storage facilities. Its 1.6-million-acre-foot capacity was down by 20 percent. Through the headset he congratulated himself, "Not bad for the thirteenth year of a drought."

We flew up dam after dam, popping over massive impoundments artfully packed into canyons and gaps, most built in the early twentieth century. As the canyon walls flew by, we saw cliff dwellings, half-crumbled fortifications of mortar and rock tucked into caves and natural shelters. They dated to Hohokam times. In the declining shallows of one reservoir, Ester pointed out a checkerboard pattern visible just below the surface of the blue-green water, the submerged ruins of a twelfth-century pueblo.

"Ironic, huh?" Ester said over the cockpit noise. "I don't think they could have ever imagined this."

The lobby of the Salt River Project's headquarters contains Hohokam dioramas of little people with tiny baskets and pots tending their water systems. "You'll find a certain pride here in the Hohokam, for the similarity of our tasks and adaptations," Ester told me.

He pushed open a pair of doors into the Command Room, which looked like the bridge of a starship. Railings led down through curved desks where two people sat in front of keyboards and batteries of computer screens running water through the city. The front of the room was a curved map, 180 degrees floor to ceiling, the entire city pulsing with lights indicating where water was moving and where it wasn't; municipal treatment plants, sprinkler systems, agricultural customers, and down-water ditches where people holding senior water rights are still flood irrigating the orange trees in their front yards.

"Two water masters run the show," Ester said, introducing me to Maggie Cherner on the left, operating the north side of the city, and Pete Cady on the right, working the south side. They looked up with a hello and quickly went back to finger-punching number pads and keyboards. They were smoothing out hourly surpluses and deficits within the system, while around them ran electronic charts and graphs. Larger screens hung to both sides of the big map relaying squadrons of numbers, headgate flows, and pipeline pressures.

"It's a human-run system," I said, surprised.

I had assumed the city was run by computer, and seeing that it wasn't was actually a relief—two live people at the heart of Phoenix's water supply, something organic still driving the system.

Maggie looked up and said, "A computer doesn't know who these people are, or really what their priorities are."

Looking over Pete's shoulder, I asked, "What's all the red?"

Pete tapped his various screens with the butt of a pencil, saying, "These are where I'm not meeting demand."

"And over there?" I pointed at another screen. "The same," he answered. "All the red you see is a problem." Ester, who was leaning against a chair, explained, "Craig's interested in the Hohokam." Pete laughed and said, "We've talked about that before. The parallels with what we're doing here and what the Hohokam did."

Maggie looked up from her monitors. "The Hohokam had to have had water masters back then, people keeping an eye on downstream users, making decisions about how to allocate resources. They couldn't have done it without some kind of cultural structure."

"And he's the rain priest," Pete said as he rolled his chair back from his desk thumbing at Ester.

Ester laughed, opening his arms to accept the title.

"This is a different civilization, though," Pete continued. "The Hohokam had immediate consequences, while I don't think we realize what is happening until it is too late. It's like the invincibility of a teenager. We're a young civilization."

Maggie said, "If we were around in Hohokam days and you asked us how much water was available, just to let you know, everybody, it's looking tight on the supply end."

"Write it down," Ester said to me. "People should know."

As I walked down this wastewater tube crisscrossed through shallow urban infrastructure, my mind wandered, brought back from the monotony of footsteps whenever the pocket mouse appeared at the farthest end of my light. Finally I saw it hop out of a hole into the next chamber. As I entered the concrete enclosure, I scanned with my light, no cockroaches this time. The pocket mouse was gone, escaped through another selection of passageways. I lifted the circle of my light until it landed on a colorful split-faced mask a few feet tall.

"There you are," I said.

The silver-gray concrete space smelled like a dirty basement and was the size of an inexplicably tall walk-in closet. Keith hadn't found it yet, otherwise he'd have been waiting, perhaps standing in total darkness contemplating the merits of suicide. I wondered where he was.

I stepped beneath the mask. This was definitely the one. Keith's son's photograph showed the same otherworldly kachina, eyes made into narrow slits. This wasn't spray paint. Someone had brought brushes and cans down here. It had feathers painted on, and dancers, and was filled with flames around the bottom, dripping with . . . was that blood?

The Hopi have stories of worlds destroyed before ours. They say we are in the Fourth World right now, and the three worlds before us fell to fire, flood, monsters, and so on. Getting from one world to the next is the trick. In Hopi legend it often involves climbing (or swimming in the case of flood) through a hole in the sky, which becomes the passage into this world.

A light flashed out of one of the tunnels. I recognized the sound of Keith's flip-flops approaching.

"Mr. Childs, I presume," he said as he emerged.

"Mr. Keith," I said, stepping back to reveal the mask.

His headlamp lifted to it. "Well, sweet baby Jesus, we found it."

"It's way back here," I said.

We weren't that far in, maybe an hour. It just felt long, as time in caves tends to. Like two men at an art show, we moved closer and back. We remarked on the way the mask had been split by black lines, not the classic Native quadripartite division representing four sacred directions, but instead a three-way split, perhaps patronage of the Trinity, some Christian influence.

"It's bigger than I thought."

"The colors are brighter."

"I wouldn't hang it with a Chiago or Begay."

"It's some dope-smoking kid having kachina visions in the bowels of the city."

Ten minutes was how long our curiosity lasted in this one concrete box. Keith and I weren't here to do forensics. We just wanted to see the thing and return to the world outside. As we headed back one behind the next, our shadows opened into one passage and another. Concrete burrows grew in size to handle outflow, the last pair big enough you could park a bus inside if not for all the hunks of concrete and shopping carts mangled together. An ooze of car-wash water and street effluent ran down the slimy middle, blue-green algae about the only thing that would take hold in here. We emerged through the final concrete mouth into a nighttime sky humming with traffic. The exit spilled its rainbow-colored water into construction debris and old bicycles wrapped around mesquites and flares of desert broom, the Fourth World, I presumed.

Canned goods, candles, batteries, plenty of fresh water—you want to have some basics arranged ahead of time, just in case. It's not paranoid. Just realistic. You never know what is going to happen. The U.S. Federal Emergency Management Agency recommends keeping a disaster-preparedness kit that includes a three-day supply of whatever it is you absolutely need.

In the summer of 2011, a public utility worker doing maintenance on high-tension wires in Yuma, Arizona, accidentally shut down electricity to four million people in Southern California. A minor mistake, and hundreds of miles away blackouts cascaded into San Diego, traffic gridlocked because signals were down as schools and businesses closed, and air traffic was grounded. One small thing. What if something bigger happened? What would you grab while running out the door?

> Headlamp
> Metal water bottle
> Fleece hoodie
> Lighter
> A good knife or multitool
> Good tweezers
> Bandannas
> Iodine
> Parachute cord
> Raincoat

That's my list, at least. Food you can usually find, a lizard on a stick, scraps at the bottom of a grocery store Dumpster, but these other items, they are civilization in a bundle, the things you might have trouble getting your hands on if the bottom dropped out. But what is it, really, we think to prepare ourselves for? Is it the one in a million, the super-volcano eruption, asteroid impact, reversal of ocean currents, barrage of nuclear missiles—an event that sends us back to the Stone Age where we may someday puzzle over our own ruins?

I used to make plans to escape the apocalypse. In high school I had a scenario cooked up. As soon as everything went to hell, I was jacking my dad's truck and his water jugs, grabbing my girl-friend, and escaping with her into the highlands east of Phoenix. Armed with my dad's guns and supplied with the necessary basics, we would get far enough to eat cottontails, drink from springs, and pluck ripe cactus fruit come June. We'd live happily ever after in the postapocalyptic world.

I never told my girlfriend, who was a great kisser, that she starred in my own personal doomsday movie. It is best to keep some things to yourself. At the time, I kind of wanted it to end anyway. Not that the malaise of suburban Phoenix life was that bad; I just reasoned if civilization was going to fall, we should get it over with and see what's on the other side. Now a bit older and more possessed by my own life, I don't feel quite the same. It is different now to think of the end and the death it brings, the children buried in the ground.

I once came upon a survival cache in the desert of southern Utah about twenty miles from the nearest paved road. I was looking behind a boulder for shade when I noticed somebody had already been there. Stocks of bullets and fishing lures had been placed along with stacks of folded blankets. I crawled into the space behind the boulder and found an assortment of aluminized space blankets half buried in dust. On a pocket chain was a ferrocerium rod and a fire striker. Matches in a tea tin were strike anywhere and stacked heads up. Fishing rods, salt tablets, twine. Sleeping bag, sweaters, tarps. Wondering what end this person had been hoping to endure, I lifted out the top layer of artifacts, dusting them off one at a time before placing them back. Was this an escape from nuclear warfare or government takeover? Maybe aliens were coming.

I found glass baby-food jars cleaned out and stocked with seeds:

corn, squash, beans, pumpkin. This was serious, a complete renewal. I did not open them, not wanting to let in outside air or moisture that would ruin these well-laid plans. What else was beneath the dust, tucked under this boulder, that I wasn't willing to dig for? Perhaps rifles and guns in PVC, barrels sealed with wax, or down there somewhere a five-gallon bucket of red wheat or barley, a piece of dry ice thrown in when it was packed to sterilize the contents and burp out the last air. Perhaps there were bottles of alcohol, gin or vodka, the kind of thing you could trade in a postapocalyptic world—lasts forever, sterilizes wounds, and useful for drinking if you have to saw off your own leg at some unfortunate moment.

So here is one view of the end of the world, I thought. Run for the hills and survive. Hope you have what it takes. And what of the rest of the world? As I rotated baby-food bottles in my hand, letting the seeds roll around themselves, I felt relieved this cache had remained unneeded.

Some archaeologists believe that ancient Phoenix was part of a much larger cultural system, a dusty outlier and trading frontier tied to what was then the Maya core of Mesoamerica two thousand miles away. Early Phoenix had ball courts similar to the Maya, and imported tropical bird feathers, and copper trinkets have been unearthed, suggesting a connection with southern regions.

If you want to see a much larger, legendary fall, look to ancient Mesoamerica. The glorious, pyramid-strewn heyday of the classic Maya appears to have been brought to a flaming halt a few centuries before the Hohokam collapse, as if what happened in Phoenix was a ripple effect, the result of a much larger failure, not just local drought and unrest.

Working primarily in limestone, the lowland Maya a couple thousand years ago engaged in monumental architecture decorated with stonework as finely executed as any dancing Ganesh or Akkadian relief. They left behind ornate and monumental cities for the jungle to swallow.

Quiriguá is one of those cities. A Phoenix of its time, it was midsized and within commuter range of the much larger and more influential city of Copán (think Los Angeles). What remains of Quiriguá today is a seven-hundred-acre patch of ruins and rocks, a stair-stepped acropolis leading down to a procession of carved monoliths in southeastern Guatemala. A U.S. fruit company set the plot aside in 1910, preserving a small portion of this lost city. Vegetation has been cleared away to reveal an open-air pre-Columbian sculpture gallery full of slabs of upright sandstone known as stelae, each delicately carved into curled and sweeping images—glyphs, stories, animals, dates, kings, wars, captives. The rocks had been etched so deeply that they looked like creatures in themselves, as if they would rise to their haunches and leap away.

In the early 1930s, Aldous Huxley visited these same stelae. He wrote, "And there they still stood, obscurely commemorating man's triumph over time and matter and the triumph of time and matter over man."

Quiriguá had been positioned along key trade routes, nestled into a broad region of Maya kingdoms that then existed within a larger network of monumental cities serving all of Mesoamerica. When you fly over Belize and Guatemala, you get a sense of the scale. The jungle is bulged into mounds, hundreds upon hundreds of them, each a former city, a cluster of high temples buried in vegetation. Something big was happening here, thousands of years of cultural development and conspicuous architecture becoming more colossal by the century until pyramids taller than those in Egypt rose from the ground. All you see now are mysterious ruins, as if the wave of a hand ended it all. From a distance of more than a thousand years, it feels as if this were just another time smoothed over, a moment lost and nearly forgotten. But look closer, and the end becomes a world in itself.

Consider Stela E, an erect 130,000-pound tower of stone carved in amazing detail, feathers overlapping down its edges while panels up the middle portray events, gods, people, names, and dates. It wears a grass-roof cap and is fenced off to protect it from further decay and admiring hands. Placed here later in the eighth century, Stela E was the largest work of its kind in the Maya world. Its glyphs show a dynastic sequence, king upon king all the way to the tapering end. It was erected just before the collapse, at that glorious peak of the ancient Maya civilization.

The monstrous city of Teotihuacán about seven hundred miles north of Quiriguá—occupied by Maya, Nahua, Zapotec, Otomi, and Mixtec people—had fallen two hundred years earlier during a mid-sixth-century drought. Teotihuacán, which once boasted 200,000 residents, appears to have succumbed to an uprising from within when the elite centers of the city were burned. This burning pattern would later become a hallmark of collapses across the Americas, elite architecture often the first to go.

Arguments for why the Maya went from peak to fall tend to land in two categories: ecological and non-ecological. On the ecological side are the familiar suspects of drought, disease, and climate change. On the non-ecological side are warfare, overpopulation, political failure. The fact that the two have been separated is amusing, as if drought had no relation to warfare, or climate change and overpopulation did not mutually pack a punch. Even now, in modern history, drought periods are statistically related to heightened conflicts. When the land is strained, so are the people. Bring it on at a bad time and you have the basic ingredients of collapse.

The most sensible answer to how the Maya world came apart between the eighth and the tenth centuries A.D. is all of the above. Everything rose to a head, and the center could no longer hold. The issue, ultimately, was carrying capacity. The system itself—the society, the landscape, and their interwoven pieces—could no longer maintain the Maya city-state network as it was. Increasing social complexity and conspicuous consumption ran smack into over-allocated resources in a time of both environmental degradation and significant drought. Regions were denuded of their summer-dry jungles for wood and timbers. Malnutrition entered human skeletons. The simplest way to describe what happened at that point is system failure. Everything went wrong. Elites had risen too far above commoners, who in turn far outnumbered elites. Intensive agriculture exhausted the land, while plant diseases common to mono-cropping were spreading.

I can paint as ugly a picture as you want of the classic Maya collapse, and certainly the worst of it was true, but even for the Maya it was not a snap of the fingers. Not every Maya city came down in the same way or at the same time. Some were vacated centuries apart. Others actually grew during the worst of times.

The drought that occupied the region during that period appears to have been patchy, inconsistent from place to place, shifting by the year or season. It was similar to what is happening in the Southwest right now—drought centers moving between Mexico, Colorado, Texas, and New Mexico so that one region is hit right after the next, while the relief of rain still occasionally falls. The collapse could not have been attributed to drought alone, though excessive aridity may have played an important, backbreaking role. Ultimately, the Maya fall appears to have involved the disintegration of political structures and elite hierarchies, along with the dispersal and collapse of urban populations. The construction of monumental architecture came to a halt in many places, and by the close of A.D. 1050 this jungle civilization was more or less over.

As I strolled between carved towers of Quiriguá, I thought of how the Maya must have watched this moment fly past them, just as we watch our history fly now. They must have thought themselves indestructible at their height. Even this smaller, secondary city of Quiriguá at its peak got up the nerve to capture the king of Copán and ritually behead him, his blood given to the gods. It was all written in stone, ancient alliances and separations between cities recorded in glyphs rising overhead, tribute to a once great future.

After Stela E was raised, commemorating the city's history, a few more stones went up, but they were of inferior quality. Copán was soon to fall, as were the other great cities, El Perú, the Mirador Basin, Tikal. Quiriguá was not as big, lasted a little longer than many, but ultimately it, too, faced the same fate: total abandonment. That was followed by the grasping fingers of ceiba trees and drooping vines, which within several centuries turned the city to mere bumps in the jungle, to be revealed in some future time by machetes and chain saws.

Through eastern Guatemala, Regan and I traveled from one Maya ruin to the next seeking out abandoned temples and half-consumed acropolises. We were apocalypse voyeurs, worshippers of the decline. After visiting Quiriguá, we continued in the Petén region, a humid and primordial basin that bears the ruins of Tikal, one of the most powerful of these ancient cities. Two of many monkey-like tourists, we climbed Tikal's temples, breaking through the jungle canopy into open sky. Getting on top of one of these temples, you feel as if it would not have been so bad being sacrificed to the gods back in the day, sent marching up limestone steps to this expansive view where you would have marveled at the vista in heart-throbbing earnest before your chest was split wide open and your life was sacrificed so the world could keep spinning.

At night, the darkness around this fallen city is profound, and the sound of clicking, buzzing insects fills the air. What had once been lit by countless fires, the largest city-state of the Maya world, was now an impenetrable black.

At night after the park had closed, I sat with a guard, his rifle resting across his lap. A bonfire had been set between two massive limestone temples. Firelight slashed through the surrounding jungle, where darts and shafts of warm orange ghosts danced between ceiba trees. A ceremony was taking place tonight, an old Maya fire ritual that would supposedly help heal the world. People swayed around the fire, their chants rising into the night as they stepped toward the flames and away. They seemed drowsy, heads moving back and forth, mumbling. Light washed up and down a pair of steep temples to either side of the fire, their heights black against the sky, twin crests too tall for firelight to reach.

Regan had somehow gotten herself invited as one of the worshippers. Maybe it was her good Spanish or her long, dark hair and Asian features that made her look a little Maya, enough to get her through the guards. She was given a bundle of candles and a corner to stand on and told she would be part of the ritual.

The Maya fire ceremony is about renewal, asking permission of the gods for us to stay here, and breaking old habits and bonds that don't work anymore. At least that was how the shaman had explained it to me beforehand. The ceremony was for spring equinox, a gathering of about twenty Maya and Hispanic worshippers and Regan. They were led by a Maya shaman preaching balance and global harmony. Before the ceremony he and I had spoken briefly, and in clearly enunciated Spanish he told me that the beautiful world that we know is made by symmetry. Without natural balance, he said, everything would fly into chaos. This applies to everything from civilization to the human mind to the entire planet.

"We are out of balance, and this is a dangerous time," he had said. "Everything must be done to seek harmony."

It seemed ironic he would come here to do it, a place that once went down in flames. But who could argue with his sentiment? Balance couldn't hurt. In a not entirely stable world such as our own, we needed all we could get. What should he seek instead? Discontent? Cacophony? Chaos? Such things come naturally; entropy, second law of thermodynamics—everything crumbles to pieces. The fall is easy. Congruity and accord, those take work.

As with most of the old cosmologies, early Maya saw humans as having a direct connection with creation. The world begins over and over and requires sustenance, ritual. Sacrifice was required if the world was to continue.

The guard I sat next to rested against the buttressing trunk of a ceiba tree. The many trees surrounding us rose to where the jungle canopy opened into a marvelous field of stars. Looking up, I made an offhand comment, saying maybe the world would really end if they didn't dance for it.

The guard was dressed in a light work coat, his face pumpkin orange in firelight. He laughed, gesturing around at the ruins with his rifle. He said, *mira*, "look," the world has already ended.

Six other guards were within reach of the firelight. Two stood with rifles over their shoulders near a carved stone nearly as tall as their heads. A few others smoked cigarettes beneath the shadowy ruins of the North Acropolis. One sat alone on the lower step of the Temple of the Jaguar, his face lapped with the warm glow. Around them was the evidence in ruined temples of what happened here: termination after more than a thousand continuous years of urban ascension.

Our impatience for the end sometimes seems palpable. Apocalyptic prophecies are countless, one doomsday for every beating heart. An excited anticipation for civilization to slip down the drain keeps some enrapt, others religiously paralyzed. We pay good money to see it in movies. There is always a new date, a fresh prophecy to pucker our expectations.

The most recent prognostication I'd heard was for the year 2012, a doomsday prophecy supposedly generated from the pre-Columbian Maya, who were not usually in the business of predicting end-times (that's more Judeo-Christian). The prophecy came from the translation of an ancient inscription found on a single stone tablet uncovered during road construction in southern Mexico in the 1960s. It was one of countless glyph-covered stones and calendar references found throughout ancient Mesoamerica, but this one seemed to have enticing references to a dire and singular future that would come to pass when the present Maya Long Count calendar cycle comes to an end. Interpreting various inscriptions, the Yale anthropologist Michael Coe wrote in 1966, "Armageddon would overtake the degenerate peoples of the world and all creation on the final day." It sounded downright biblical. How could one resist, especially if you were poised at the edge of boredom, drumming your fingers, waiting for the apocalypse to come?

In 2011, a formal response to this supposed Maya prophecy was offered by Grand Elder Wandering Wolf, Cirilo Perez Oxla, representing the National Council of Elders Maya, Xinca, and Garifuna tribes. The letter politely explained that conditions on the earth were rapidly becoming worse, and if something is not done to curb the impact of civilization soon, we are all going to pay the price. "We will see hunger and drought," the elder wrote. "Plagues will invade the fields and affect the agriculture; new illnesses will appear and will be difficult to cure. The sun rays are getting stronger and stronger as time goes by."

It made you want to say, yes, we've noticed.

"If we do not change," the Grand Elder added, "few will be the ones to survive." At the bottom of the page was a footnote: "Contrary to popular belief the living elders of the Maya do not agree that December 21, 2012, is the end of their calendar."

So much for that prediction.

But when it comes to relatively swift and tragic endings, the Maya could be considered experts. They have felt the sharp stick many times. You could say their end came with the fall of their largest city-states during sociopolitical decline nearly a thousand years ago when the elite palaces of Tikal were burned. Or did it come with the European-introduced epidemics that followed, wiping out most of the sixteenth-century Maya population in a macabre environment of lesions and corpses? It could have been when the last standing Maya cities were taken down in fire and blood, the Spanish conquistadores bringing unfathomable warfare and genocidal slavery. You could even say it was much later, in the twentieth century, when paramilitary death squads flooded these same jungles, their atrocities unspeakable.

And yet Maya culture still exists, with a population of at least six million living in Central America, speaking twenty-one different dialects. If you ask a Maya when it all ended, when they ceased to exist, you would discover how farcical the question really is. There was no end, no final switch thrown to shut off the lights. You might say that civilization as a whole has never fallen. Our global enterprise has not paused since the first large-scale human settlements appeared six thousand years ago in Southeast Asia. Parts of it have gone under, but at the same time others have risen. The baton keeps being passed ahead. This is the rule of the phoenix, the myth of eternal return.

As much as the guard with his rifle across his lap was right that it all had ended, there was another story happening in Tikal on this night, more than settling destruction. Acolytes around the fire, my wife included, instead of preparing for the world to end were calling for it to continue.

The guard shrugged and told me that at least the ceremony was a break from the usual, his nights long and dark as he patrolled these grounds. To him, this civilization had completely ended. It was stone enwreathed by roots and vines, silent beneath the click and scratch of insects, a city long fallen. But out of it came tonight's fire. He and I sat and watched as one after the next people approached the fire along chalk lines—explained to me as the four quadrants of the universe, the holy Native American quintessence. They tossed in handfuls of colored candles. Wax hissed and popped in the blaze. Runners brought more wood to keep the flames high, additions of planks and two-by-fours elevating the light until it grazed the temple crests 180 feet above us.

Hundreds of candles went into the fire that night, each carrying an intention for cycles to continue, familiar processes of life breathed into them before they were thrown into the flames. Wax ran out like lava until all hands were empty. The ceremony rose, people wavered. Energy encircled the shaman. He became a charismatic figure, chanting and lifting his arms in trembling, ritualistic gestures. As if swept up, he moved from one person to the next, cutting across their throats and hearts with the swift blade of his hand, what to me looked like a show of sacrifice. Both the guard and I watched as his black form moved against flames, a vacancy, as if we were able to see through him to someplace else.

5

Cold Returns

In the space between chaos and shape there was another chance.

—Jeanette Winterson

WEST COAST, GREENLAND

IN THE FUTURE there will be ice. Lots of ice. Sea levels will drop, and the ice will return, water shifting from ocean to land, turning Florida back into a swollen thumb and putting up a thousand-mile-wide bridge between Siberia and North America. Whether this happens in thousands of years or millions, it will happen. Ice always returns. Glaciers grow like roots around the ball of the earth, ice sheets spreading across the brows of continents, burying familiar landscapes miles deep. Occasionally, the earth falls into year-round winter that can last up to 100,000 years, and a time or two, hundreds of millions of years ago, the whole planet has gone under, ice believed to have extended from pole to pole making this a completely white earth, seas brought to their low point.

If this future has humans in it, I imagine them as warren-building icemen. Icemen running across vast expanses of nothing but white, footfalls breaking through to the rhythm of their breath.

I hung back a little as two of these modern icemen hurried across crusty spring snow in western Greenland, hunching along the edge of a rise so they wouldn't be seen by a muskox on the other side. They were surrounded by glaciers, not the steep, crackling, moulin-riddled glaciers you get at lower mountainous latitudes like Patagonia or the Alps, but robust arms of valley-crossing ice flows. Healthy ice. It crawled jagged into the valley floors, wild, splintered teeth still advancing from winter, aggregate boulders of ice tumbled in front of each other leading us into the month of May.

The men's faces were unshaven, their eyes bright and curious. One kept looking back at me, flashing a shit-eating grin. He knew we weren't supposed to be doing this. On the other side of the rise, the lone woolly muskox was pawing and huffing down to dry tundra grass beneath the snow. We were trying to see how close we could get to it.

There were no other animals. No trees. Nothing but erratic, black-striped boulders poking through the spring thaw. East of here, the ice sheet consumed everything, taking up a space 1,500 miles north to south, 680 miles east to west, and almost 2 miles thick at its deepest (and highest) point. We were at the Greenland ice sheet's snaking, probing boundary. The two icemen were a pair of French snow physicists waiting for their flight to a research station on the ice sheet. This was a busy time of year, researchers departing from this coastal logistics base at Kangerlussuaq to start their seasons at remote camps and facilities in the void of Greenland's ice-capped interior. I was waiting for my own flight, heading to one of the smaller outposts, a seven-person camp about two hundred miles from here. The only way for us to get to any of these places was by air, and weather had been making landings hard, most flights recently aborted in sheeting wind and limited visibility. Sometimes you wait weeks for the window. Feeling cooped up in the narrow, ship-like halls at Kangerlussuaq, we had rented a pickup truck from a local Inuit man and driven out an ice road looking for wildlife.

It was a slightly overcast day, warm enough you could get away with light gloves and a wool hat, nothing like the weather on the nearby ice sheet, where cold begot cold and wind did the same. Down near the coast, the freezing winds off the ice sheet pushed into warmer marine air, where sometimes you'd catch a jet of cold, a filament from the interior. We darted like scrawny wolves over the rise into the muskox's blind spot. None of us knew the first thing about muskoxen. I had only heard they were notoriously grumpy and could turn on a dime. Beneath its sleek rug of guard hairs, the solitary bull looked to weigh about seven hundred pounds. Most of it was one big muscular shoulder, as if the rest, the face, legs, and hind, had been an anatomical afterthought. I knew enough to stay about fifty feet behind and slightly uphill of both the Frenchmen. If anyone was to be trampled and crushed by an iron-horned creature, it would be they, and I'd be running for my life in the opposite direction.

One of the men was a tall, innocent-looking graduate student who'd never been to a polar region and didn't really know what he was getting himself into. The other was a principal investigator, head of multiple research teams, a well-published man in his forties. A stocky, apelike fellow who travels the world on ice research, he had told me that being stationed in the Greenland interior was his least favorite, not because of any dearth of scientific material, but because it was so severe and featureless. "There's no relief once you're on it," he said. "To me it's the most dreadful place on earth."

I was going because I wanted to witness the earth in its deep winter phase, the result of many thousands of years of hard winters and short summers. At the tail end of a polar winter, the Ice Age seemed far from over. The specter of summer melt had yet to begin, the melt region known as the ablation zone moving deeper every year. But for now, everything was still frozen, and I was eager to catch my flight toward the center. The head French physicist had told me that the visual deprivation of the ice sheet can become maddening. "It's all in the details out there," he said. "It's just a faint whisper. You have to listen hard if you want to hear anything."

That's why he wanted to see the muskox up close, one last picture in his head before he took off for this blank spot where the world is reduced to equipment, a handful of too-familiar faces, and a great emptiness. The heavy, sleek-haired animal wandered the snow, pawing, nipping. It knew we were here. The Frenchmen thought it was oblivious to us, easy prey, but it was just taking us in stride as it ambled along the backside of a glaciated ridge, not looking directly back at us. This was as close as we were going to get, maybe forty feet. The pair dropped behind a gray glacial erratic the size of an office desk. I found my own boulder and came to my knees. We sneaked up our heads and peered at the bull. It abruptly stopped eating and spun to face us, not liking this kind of behavior. With a frustrated stamp of its baseball-sized hooves, it snorted and butted with horns that looked like a single battering ram, curled at the tips to hook your leg and send you flying. They'll throw wolves to each other, these muskoxen, stomping them into the herd.

The muskox gave us its warning display and turned, galloping the other way with surprising speed. We stood from behind our respective boulders to watch its shag flap as it ran. As I watched it go, it was easy to think I was in another time. A balance had tipped the other way. We were standing somewhere in northern Appalachia, or the edge of Iowa, or even London in an era when everything to the north was uninhabitable, consumed by ice, and locked into weather patterns you'd be hard-pressed to survive for one night. Along with such conditions an ice age is host to other global unpleasantries. Cold air carries less moisture than warm, resulting in widespread drought. Deserts tend to be much larger during glacial periods, and climates are more unsettled, prone to heat waves and deep-cold plunges out of nowhere. Seas in those times are populated by fleets of icebergs, those coming off North America floating across the Atlantic and grounding on the shores of Portugal. This is not necessarily the friendliest earth to our own kind, dry, icy, and populated with large tramplers and meat eaters living at crowded glacial margins. I was looking for the core of such a glacial time, winter earth as modern humanity has never seen it.

On an absolutely clear-sky morning, a twin-engine ski plane shook and buffeted against oncoming wind. It crossed over what looked to be an infinite expanse of nothing whatsoever, engine drone fighting with the wind. Herringbone fractures and deep, blue chasms passed below in the ice sheet. Otherwise, it was nothing but a blank white canvas.

A space had opened between storms, and I picked up this flight to camp. I was crammed in the back of the cargo hold with gear and supplies stacked nearly to the ceiling, crates of sensors and electronics, boxes of food, a wooden gear sled, and many wine bottles taped up so they wouldn't shatter en route. The other passenger was José Rial, a charcoal-gray-haired chaos researcher and climate change scholar out of the University of North Carolina who was born in Spain and raised in Venezuela. We were both buckled into jump seats against the rear bulkhead. Out of his rumpled coat, José pulled two small, plastic airplane bottles of rum.

"For the flight," he said in a gravid accent.

He handed one bottle to me. We cracked them open and clicked plastic necks together to toast to our journey into the abyss. José shot back a quarter of the bottle and ducked his head into his oval window to see the ceaseless, white expanse roll out beneath us.

"The ice sheet in all its glory," he said. He twirled his hand in the air, adding, "As Darwin would say, so much beauty for so little purpose."

José seemed to me like some old, cunning Latin poet, half his sentences starting with oration. Continuing to watch the ice sheet pass, its surface fanned and feathered with blowing snow, José told me he was about to start his sixth season working out of this camp. He was planting seismic sensors across the ice, his ear to the ground listening for swarms of rumbles and ice quakes. He wanted to know how this major earth feature of Greenland functions, what secrets it tells in all of its background noises. He knocked back the rest of his rum and tucked the empty bottle into his coat. He folded his arms, slumped into his own chin, and closed his eyes.

I had gotten in with a research team led by the ice-climate scientist Konrad Steffen, who goes by the name of Koni. José was one of Koni's field partners. There were also two white-bread, Midwestern NASA boys who'd come in behind us, a sailor-mouthed French graduate student from the University of Colorado, and a young, wiry mule of a man from Tenerife here to help with shoveling and other manual labor. José and I were arriving first to open the camp, which had been closed for winter.

As a scientist, José was a sort of ragpicker, looking through strings of data other scientists throw away or would never touch, what they call white noise, anomalous behavior. He is interested in what *cannot* be predicted. He told me, "Seeing a pattern is mostly in your mind; it's like reading tea leaves." While many scientists look for the most robust patterns in natural systems, José hunts in the background for less well-explained indications that might have been missed. Though he is suspicious of interpreting patterns, he combs chaos looking for them just to point them out, listening to static for a sign of conformity, the presence of a secret code. On the ice sheet, one of the planet's self-generating refrigerators, he records creaks and snaps, the sounds of mechanical processes inside this massive, singular glacier. Some are surprisingly uniform, rumblings that last forty-five minutes starting with calving ice from the outside edge and propagating miles into the ice sheet. Media reports have taken these findings as signs of imminent doom, the Greenland ice sheet coming apart, its physical structure failing. Some scientists have said these sounds suggest fractures are coming faster than ever before. José calls this kind of talk alarmist. He thinks the sounds are happening all the time, a natural phenomenon. For José, they are ciphers, delicate snaps and grumbles that express the inner workings of an ice sheet in progress, not necessarily a brazen catastrophe. He looks for subtle signals that might give us keys we need to understand how this often temperamental planet actually functions, not just the obvious driving forces such as orbital variations and carbon emissions. He knows that the actual future is the one we never expect.

Props revved louder as we descended through a stiff surface wind, and as we dropped, the relief of the ice sheet changed. Its flatness curved into waves and troughs like the long, gentle rolls of waves in a giant sea. Tides? I wondered. Great sea ripples? The plane swerved into the wind, coming down over a basin so gradually sloped it'd be hard to get a pencil rolling down it. The shadow of the plane rose to us, and with hardly a bump we dropped onto skis and taxied up to a disheveled pair of red wind-battered Quonset tents. As the left prop wound down, I pulled on a parka big enough to sleep in. I tugged on mitts, climbed over a stack of yellow torpedo-like propane canisters strapped to the floor, and unlatched the door, which flew open in the wind. The oily rag smell of the cabin was stripped away by freeze-dried air. It smelled like oxygen, rarefied, stinging the rims of my nostrils.

It seemed I'd waited all my life for this moment, stepping onto five thousand vertical feet of ice, 684,000 cubic miles of contiguous frozen water. I had practically rehearsed this climb down the ladder, one small step for a man and so forth. The right prop was still running, ready for departure. The pilots had other pickups and drops to make in this gap between storms. They wanted us to hurry. I turned backward and climbed down. Searching for the next step with big, clumsy moon boots, I missed my mark and pitched over backward, leaving the plane much faster than I had planned. I landed hard, hitting the snow like a sack of flour. That is how I arrived on the Greenland ice cap, on my back sprawled like a bug.

José appeared through the door and waved at me to get out of his way. I pushed up to my knees, and plates of snow broke apart, skittering away in the wind. The sun hung bright and low, and I blocked it with one large mitt.

The surface was not solid ice, as you might expect, but snow. I tromped around to get my bearings, but there weren't any bearings to get. Sleek, low-profile snowdrifts folded over each other, nearly sanded flat by a constant, sugar-sprayed wind. I hooked the door to a strut so the wind would stop banging it closed.

"Let's get this plane unloaded," José said.

The pilots moved fast, sliding boxes down to us, and we waddled them off, stacking them out past the plane's wing. It was a five-minute unload, about a thousand pounds of gear. The pilots patted our shoulders, gave a thumbs-up, and climbed back in the cockpit. The left propeller kicked up a loud drone as I backed away. The plane pushed ahead and skimmed over the wind. It rose into a dry turquoise sky, then it banked south, the sound of its props replaced by wind.

José marched toward the two red work tents as I gave the place a slow 360. Any sense of distance was impossible to gauge. Cold and dense, the wind was racing down from higher ice in the east. I turned to see José stumbling up the high fin of a snowdrift draped around the side of the kitchen tent. Shouting over the wind, he said, "What the hell happened here?"

Now that I had a better look, I could see an open mouth in the lee of the tent where the door had sprung open during months of winter abandonment. Snow had poured in nearly up to the top of the doorjamb. The platform was twisted. Built into the ice, it had shifted around and camp had come apart.

"Oh, this is bad, very bad," José moaned.

As I looked around camp, more damage was obvious. Metal towers had snapped, half protruding from the sculpted surface. These were not from shifting ice. Storms had broken them. Pieces of plywood stuck out at odd angles. Solar panels were twenty feet from where they started, now facedown and barely exposed. Snowmobiles should have been parked on a wooden platform held up by metal posts anchored thirty feet into the ice. The snowmobiles were gone, along with the platform and the metal posts.

"A third of the camp is gone," José said, pacing around, pawing at his hood to angle it against the wind.

"Where are all the snowmobiles?" I shouted back at him, wondering, too, there was a radio, right?

José waved his glove at the ice sheet. "I don't know, somewhere down there."

I looked under my feet. Down where? There was five thousand feet of "down there" beneath me.

José climbed the snow cornice into the door, cleared a hole, and slid down inside, boots last to go. I climbed up behind him and cupped my eyes. José was post-holed knee-deep inside a space partially buried. Everything had been left in order, rice cooker under a mound of snow, wineglasses hanging from a rack like white bells.

José said, "This is going to take some work."

"Is the platform stable?" I asked. The supports looked bent from the outside.

José jumped up and down, thudding the snowcapped floor. The structure did not collapse.

"It feels sound enough," he said.

Let me reiterate. José is a chaos researcher. His specialty is abrupt climate change, things that could happen out of nowhere. He studies deterioration and random decay, searching for previously undetected or ignored background signals in the data, believing that between disorder and form is the real world.

When I pulled my head out of the kitchen tent to the expansive sun and wind, there was nothing to reference visually. Around the perimeter, broken weather stations and our pile of crates and propane bottles were the last material substances. Beyond them, whiteness stole the eye, brightest wedding dress I had ever looked into.

José and I pulled shovels and got to work, about fifteen hours to nightfall and we'd want this cleared out and have our sleeping tents up.

This wasn't the kind of snow you shovel off a sidewalk. As dry and packed as gypsum, it was easiest broken off in big chunks, shovel wedged to crack off blocks you carried away and dropped down the slope surrounding camp. After a couple hours, we had enough space cleared to sit, folding chairs pulled out around a circular plywood eating table. While José chiseled out boxes of food left from last season, I turned to the second tent, Koni Steffen's field office. Hammering away with the blade of my shovel, I squeezed through drifts, pried open the torqued plywood door, and slid inside, where snow had been drifting in for months. Weird ghost shapes had gathered on everything, and the ceiling hung with eerie stalactites, fine snow crystals clinging to each other like feather down. I strayed through these soft, hanging shapes, my shoulders brushing them apart. Low sun through the red enclosure gave the space a blood-colored hue, a frozen inferno, a vampire's chamber. Drifts rose from worktables, overwhelming shelves of sensors in various states of construction. In my mitt I picked up a candle burned halfway down, imagining Koni, the scientist, sitting here writing his late-night papers. Koni is one of the more trusted voices in the profession, and after thirty-five consecutive seasons in the polar Arctic—overwintering twice in a tent—he is the one you want to ask, what will happen to the ice?

I excavated Koni's desk, clearing off electrodes, circuit boards, wire strippers, and stubs of pencils. In the middle of digging, I pulled back the ruff around my hood to the sound of prop engines far away. A storm had been moving in, and we weren't sure if the next flight would make it. Was this just a flyover looking for a place to land, or were they really coming in?

José knocked on the crooked plywood door and said, "They're landing."

The same plane that had dropped us off fell through the wind and touched down. Another three from our crew climbed out. No one, I noted, tripped off the short descending ladder. First was José's nephew Nestor Rial, the guy from Tenerife, a thirty-three-year-old with the energy of a kid, jumping onto the snow with his arms wide open, laughing at the expanse surrounding him. He had been dreaming of this moment since childhood.

"This is incredible," Nestor shouted. "I need to hug someone."

I was the closest, and we crushed our parkas into each other.

Nestor was followed by Nicolas Bayou, call him Nico, the French cryosphere graduate student.

Koni was last out of the plane. Lean and bone faced, he had a thick beard neatly trimmed, and it iced up as soon as he emerged into the cold air. José explained to him what we found when we got here, describing the amount of snow we'd already cleared.

As the plane departed, Koni walked around camp, surveying the damage. In one swift move, he lit a cigarette in the crook of his shoulder. Holding it in his teeth, confidently sucking to keep the thing lit in the wind, he said, "I've seen worse."

We shoveled the rest of the day, put up sleeping tents in a snowy gale, and near eleven at night found ourselves sitting around the circular table, five men in coats with greasy pad-thai dinner plates finished, wineglasses drained. The kitchen was partly in order, snowdrifts chiseled back to the corners, floor frozen solid. Outside, wind sounded as if it were destroying the planet. The storm had settled in. Until further notice, all flights were off. The two from NASA scheduled to join us would be stranded in Kangerlussuaq.

A gray-blue light of dusk came in through one small, square Plexiglas window on the windward side of the tent. Heavy tarp-like poly fabric banged forward and back as wind sucked around the tent.

Koni was telling stories. Zurich-born and working out of the University of Colorado as the head of the Cooperative Institute for Research in Environmental Sciences, Koni had just the right Swiss composure with a controlled American swagger that made him a joy to listen to. He told of the time he and his wife hightailed it through the Hindu Kush in Afghanistan with a Soviet invasion on their heels. After a few more stories went around the table, they inevitably returned to climate before reaching Koni again, who told us that the previous winter he had appeared on MSNBC, where he had been asked to explain why temperatures that winter were suddenly warmer in Greenland than on the East Coast.

"They wanted to hear it was global warming," Koni said, his accent lengthening every vowel. "Because I wouldn't say it, they only used two minutes of the video."

He was frustrated by this. He had spent one hour in front of a green screen like a weatherman explaining ocean currents that wind up the Eastern Seaboard, lodging warmly around Greenland. He told them it was not global warming but a predictable cycle of ocean currents as the Gulf Stream fluctuates where it meets Arctic waters.

"And still it came out sounding like global warming anyway," he said.

José twirled his empty wineglass and said, "They want to know that they are doomed. They are obsessed with their own mortality."

"Bullshit," added Nico, the Frenchman who was grumbling over a glass of Saint-Vivant Armagnac. It came out sounding like *boosheet*. "They want us to feed them bullshit."

Koni uncorked another bottle of red. He tapped the neck on each of our glasses. He was happy to be here in familiar company with the wind banging outside, home again in a swan-white landscape going on forever, his true love.

José lifted his glass and said, "Happy mortals."

We all lifted glasses.

We were feeling good and healthy after a day of shoveling in a snow-blasted gale, clearing out the generator, putting up sleeping tents. I was impressed with everyone's poise at this point, sipping wine, snapping off little bits of fine chocolate Koni was passing around while a storm socked us in tight. Work would start again first thing in the morning. Finding the snow machines was a priority. When they were recovered and repaired, we'd be burning almost every hour of daylight getting out to remote sensing stations for downloads and maintenance. This camp was here to record conditions and movement on the ice, monitoring one of the more crucial and telling landscapes on the planet. What Koni had discovered in his decades of work is that big, perhaps unprecedented, changes are afoot, the loss of ice accelerating. What this means for the future, he is unsure.

The flimsy, RV-style door banged on its bungee cord, wildly slapping in a back draft. It was loud enough to silence conversation at the table. I excused myself, said I needed to pee. I really just needed to go outside. I couldn't stand hearing it around us and not seeing where we were. At the door, I put on every layer I had, three coats and a parka, wool hat beneath a hood pulled tight.

Koni eyed me as I snapped, zipped, pulled. "Be safe," he said.

Tugging on wool gloves and heavy mitts on top of them, I mumbled an affirmative through a mouthful of coyote ruff.

Opening the door was like breaking an air lock. I stepped into oblivion. Twenty below zero Celsius, according to the wall thermometer inside. Drifts were flying by, carried in the roar of katabatic winds. "Katabatic": from the Greek for going downhill. Wind starts at the ice sheet's 10,700-foot summit and races about three hundred miles past here to sea level. On its way, it picks up speed, and the only obstacle it encounters is this lone research station. Speeds of 80 miles per hour were sometimes sustained for days. I was guessing about 60 tonight.

The pee pole was a hundred yards downwind from camp along the tail-fin ridge of a snowdrift. A splintered bamboo stake had been hammered into the drift. That way we didn't end up peeing anywhere we pleased, getting crusts of yellow snow in the drinking water that we melted on the stove. Pissing in the wind has its own hazards. You don't want to come back sequined in urine crystals, makes you look like a bad polar explorer.

The northern sun was still up, though I could not tell where. Visibility was one or two hundred yards at best, interrupted by swift and blinding curls of snow, through which nothing could be seen. It was not actually snowing. Instead, this was a dry storm. I was, in fact, in a desert. The snow blowing past had started hundreds of miles from here. In a constant morphological change, the snowflakes around me were being obliterated every hundred yards or so, turned back into a gas before the intense cold compressed them again into pinpricks of newly formed snowflakes. Each would last a few seconds before gassing out once more. That way, the water was being constantly purified. You shouldn't even drink it straight. Koni had warned that there aren't enough minerals in it, the water so pure it pulls salts and electrolytes out of your stomach lining, which makes you sick. He insisted we add tea or sugar to anything we drank.

Finished there, I walked out past the pole, feeling small snow-drifts riding up the sides of my boots with every step. I didn't go far, mind you. Ten paces, maybe twelve. Camp created wind resistance, forming a snowdrift about twelve feet higher than surrounding terrain tapering out a quarter mile like a comet tail. It was a weave of dorsa, a filigreed piece of nonlinear science unfurled across the ice sheet. In the ghostly, polar half-light I could make out the plain that earlier in the day had been a blinding bone-white circle. The storm socked everything in. I thought this was perhaps the most uninhabitable place I had ever seen. Animals don't come out here, and certainly not plants. Koni told me you see a bird now and then, some wastrel lost and blown off course. Apparently, they come to camp because it is the only solid thing, the only visible thing. He doesn't bother saving them. There is nothing he can do. "They come here to die," he told me.

By morning, they are frozen solid.

I felt my face with the nylon tip of my mitt. Was this the sensation of frostbite or just the thousand-pointed prickle of snow salting my cheeks? Already Nico had earned a white patch across his right cheekbone from shoveling earlier today. Tomorrow the skin would be the color of ocher, the next day a dry and crinkled brown.

Alfred Wegener died somewhere right around here. Wegener was the German scientist who introduced the idea of continental drift in 1915, a hard sell among geologists. He laid the groundwork for plate tectonics theory and ultimately revolutionized the way we look at the surface history of the earth. Charts put his expedition in the fall of 1930 not far east of here. He was out on an Arctic weather survey, and when supplies ran low, he and a companion named Rasmus Villumsen set off on two dogsleds to reach a resupply camp. Along the way, they planned to kill dogs from one of the sleds to feed them to the remaining dogs so they could keep going. They never made it to the camp, however, and Wegener died not far from here. Villumsen buried him, crossing a pair of skies to mark his grave. Villumsen left from there and was never seen again, his remains likely drawn down into the ice, where his stretched and distorted remains will eventually be carried toward the coast and will spill into the sea embedded in an iceberg.

I looked back to make sure I was still in line with the pee pole and camp behind it, which was ranging in and out of sight through blowing snow. I pulled my hood to block the wind and rounded my back against it, making a statue of myself. A downwind drift began to form. Sleek tendrils grew out from my boots and moved west. Soon, I had my own comet tail, the snow gliding around me as if I were just another snag to smooth over.

The esteemed nineteenth-century scientist Louis Agassiz was the first to propose that parts of the earth had been wholly consumed by ice not very long ago. To many, the notion seemed preposterous. The consensus of most scientists at the time was that erratic, boulder-strewn, and ice-gouged landscapes were the result of a single apocalyptic flood. Dominant landscapes such as the fjords of Norway (or the Grand Canyon for that matter) were thought to have been formed by a single biblical event. In other words, it happened all at once as Noah boarded his ark and the rest of the world vanished beneath tempestuous waves.

Agassiz saw a more plausible if not more fantastic history. He looked at the glaciated horizons of Great Britain and determined that at one point not so long ago the entire place had been under an ice mass miles deep. He turned to broad geologic scours that had been explained as drag mark's of icebergs careening in the Great Flood and said they were actually the result of solid, continental ice grinding slowly across the land. He imagined the same for nearly all of Europe and North America, and even went out on a limb to say this happened not just once but many times, ice advancing and retreating in long, slow waves. At first, many geologists murmured and scoffed. The scale was simply unimaginable. But Agassiz imagined it for them. He laid out maps and drawings, traced courses of glaciers carved in rock.

In 1837, Agassiz wrote, "The epoch of intense cold which preceded the present creation has been only a temporary oscillation of the earth's temperature." The theory proved physically irrefutable. Using a term coined by Agassiz, we now call these oscillations ice ages, and knowledge of their existence altered the way we thought of time passing on earth. His discovery was a leap of context.

In Greenland, cores drilled out of the ice have hit bedrock thousands of feet below, and one drill site came up with spruce tree needles. There had been forests down there, a very different Greenland, actually *green*. Before the ice. But even with reams of hard data that have been accumulated since Agassiz's time, no one knows exactly what starts an ice age. The major driver appears to be changes in solar radiation reaching the earth. This is based on the way the earth swings toward and away from the sun over tens and hundreds of thousands of years, gradually warming and cooling by very slight degrees in a predictable rhythm. But computer models using these orbital variations still don't produce the sudden onsets of ice ages.

Remember Aixue Hu's findings that opening and closing the Bering Strait can turn ice ages on and off. Smaller changes on the earth's surface magnify the bigger ones, as if the world were a megaphone shouting into itself. In the past, interglacial periods similar to our modern Holocene have ended and entered a full ice age in the span of a century. It starts when summer polar temperatures are low enough that the snow doesn't melt and lasts through to the next season. For ice sheets to grow to their full extent, then, takes thousands of years, but that first century would be a doozy, and if we saw it today, global food production would plummet in the face of increasing cold.

Most living creatures don't fare well in the glacial lows that come with this cold. There is a reason people didn't cross the Bering Land Bridge into North America until the ice sheet blocking their way began to recede. Even spear-bearing, woolly-mammoth-eating badasses from the Pleistocene weren't walking across a place like this. This is global death by ice, a force as rendering and earth changing as molten lava.*

Current global warming aside, the earth has been in a long-term cooling phase with temperatures creeping downward for about sixty million years—likely due to a period of extensive mountain building, which naturally reduced atmospheric CO_2, removing insulation from the skies. Global warming may be the issue for the moment, but the bigger, multimillion-year trend is toward a cooler planet. Antarctica iced over about thirty-five million years ago, and ever since then extended ice ages have become almost commonplace, growing into a discreet rhythm of rise and fall over the last five million years. In that time, even during interglacials, ice has always remained. There has been a cold heart waiting to start up again.

All morning we dug, trying to reach the snow machines, which were buried along with their splintered platform under ten feet of

---

* It should also be noted that the presence of cold, especially in the context of a heat-trapping atmosphere such as our own, is one of the planet's saving graces. The rise of oxygen 2.3 billion years ago, known as the Great Oxygenation Event, coincided with the first identifiable ice age. Oxygen rebalanced earth's atmosphere, diluting greenhouse gases, giving the planet its first chill.

cover from winter. The snow was as compressed and thick as half-set concrete. The wind didn't help. I swore that for every shovelful I threw out of the hole, another two were blowing back in. My beard was a mask of icicles. So was Koni's. Nico, the broad-shouldered Frenchman, was doing twice all our work, throwing planks of snow out of a hole now eight feet deep. Nearby, José found a treasure box of stove parts and bungee cords four feet down and together with Nestor was exhuming it, pulling out harnesses and drill bits.

I stopped for a breath and leaned on the handle of my shovel. That was enough time for five minutes of work to cover back over, drifts climbing my boots like vines. This fresh snow cover wasn't packed down like what I just dug out and would be easy to clear away, but it was disheartening to see my work smoothed back over so rapidly. The place was relentless. But slowly we were getting somewhere, recovering lost equipment boxes and torpedo-shaped propane bottles two hundred pounds each, fallen all over one another. It felt like any scientific excavation I'd ever been involved with, uncovering tumbled remains of ancient things, only this was gear from last fall, end of the field season. Before any of the science could start, the camp needed to be secured, equipment accounted for and available.

"This is peanuts, digging out a few Ski-Doos. If you want to see hard work, a late-fall melt filled the tent with ice, and it was like a skating rink," Koni had told us this morning around coffee and dry cereal. "This, today, is nothing."

I was glad this was only snow. It would be another thing chipping out ten feet of solid ice to disentomb our snow machines and whatever fell in with them. Down on my knees clearing around the upside-down runner of a wooden sled, I stopped and looked up, hearing the sound of a plane. We were all looking up for it, shielding our faces. The shape of the plane fell in and out of view, same Twin Otter that brought the rest of us. The NASA boys were trying to land, looking for a hole.

"Not today," Koni said.

We heard the plane circle three times and then turn back for Kangerlussuaq.

I planted my shovel and climbed out. Pawing over cakes of snow I threw out, I stood with a better view and realized I had been down there for an hour without looking up. Out of the hole looked about the same as in the hole, ground-level visibility edging on whiteout. I couldn't see a damn thing. Koni was standing up top, perched on the blocky snow piles we'd created. He was smoking a cigarette as he faced east into the wind, his hands jammed into the pockets of his shell.

Koni was one of the healthiest-looking human beings I had ever seen.

"Is it always like this?" I asked him.

Koni glanced my way and lifted away his cigarette with rawboned fingers, knuckles as big and square as dice. He only took the cigarette from his mouth for as long as the bare skin of his fingers could take the cold, five seconds.

"It's often like this, yes. In the winter it's much worse, though."

Well, this sucked.

"We're not going to get anything else done in this wind," Koni said as he planted his shovel and turned for the red tents. "Coffee."

People came climbing out of their holes as if the recess bell had just rung.

But I didn't drink coffee. I carried a shovel over to clear out my tent so I could get in the door, grab a camera, take a picture of this relentless wind. Tents were every man for himself. You didn't go dig out someone else's tent. It was fruitless, becoming snowbound again within the hour. You dug yourself in and out each time, and otherwise you just let snowdrifts crawl all over everything.

I cut a U to my door, pulled the first zipper and the second, grabbed a camera, and closed up behind me. I don't know why I thought I could get a photograph of this. Maybe just for the pearl-gray sweep of the storm. I stood at a line of gear boxes not yet buried, their sharp edges and low profiles shark-finned through the wind. They marked the end of camp. Beyond them was a howling hole. Everything else in the world seemed gone, never to be recollected, my cities swallowed, all the ships at sea frozen in place, slowly crushed by the unbearable weight of it. For what I could see, there was no end to this.

I stood there wondering how I could even think to comprehend this place. What brain grasps complete emptiness? I had eleven days before a Huey helicopter was scheduled to pick me up and I would trade out for another crew member, and I feared I would not have the time to even scratch the surface between now and then. Last night, as I had lain in my tent feeling the weight of snow gathering, pressing me into a smaller space, I heard José cursing outside. In the perpetual half-light of a night storm, he had gotten up to dig out his tent. He must have been afraid of its collapsing. At least he wasn't shouting for help, and frankly I wasn't feeling too altruistic at that moment. Nobody was, because we all heard him and nobody got up. This was the simple triage of trying to get some rest. After a while, José was quiet again. He was either dead or back in his bag, either way problem solved.

How many places on earth would that thought go through your head, even if you wouldn't openly admit it? I have heard from climbers of Everest who gasp for air at towering heights that they can barely get out of their tents in the morning to save themselves. At least in Greenland we had plenty of oxygen, and Koni even brought fondue to have with chocolate and wine. If I could complain, the worst part was getting out of the bag every morning and jamming yourself into frozen clothes, or your hands burning while you unrolled toilet paper at the outdoor shitter, which was a hole in the snow behind two propped-up pieces of plywood used as a wind block. It was relatively comfortable as long as you were dressed like a bear and didn't take your mitts off for more than five seconds.

I turned back from the edge of camp and stomped my way over to the door to the kitchen tent. I cracked it open, and a wall of steam came in with me, cold air hitting warm stove air with a second round of hot coffee brewing in two moka pots. I stomped off, as greetings came in many different accents, men sitting at papers, laptops, Koni scribbling in pencil in his battered yellow notebook. This was what he meant by coffee: grant proposals worked on; scientific journal articles on climate change given a thorough peer review.

I pulled off my mitts, wool liners damp from sweat or snowmelt or both. I peeled off layers and hung them, pulled out a folding chair, dropped into it.

José looked at me over the rim of his glasses. "You were out meditating on the ice," he said.

"You think an ice age could start up again anytime soon?" I asked.

Koni laughed at the question, amused, but didn't look up. Nobody seriously talks about a coming ice age in this time of global warming, at least not serious, peer-supported scientists. Among even fringe scientists, the notion of a coming glacial period has all but been abandoned like avocado-green refrigerators, a retro idea from the 1970s. It used to be popular back during pulses of global ice growth. In the 1980s, Koni witnessed glaciers advancing in Norway, pushing through forests, splintering and knocking down trees. That hasn't happened since, and the game has changed anyway, pretty much nothing but retreat. Data sets are better understood now, fine grain of observation made exponentially finer. Computer models are in general agreement that nothing but warmer and more erratic climates are in our future. It was the erratic part that interested me. How erratic? Could we somehow swing into a new ice age?

I sat facing José, who was saying, "By its very nature, change is highly unpredictable. And nature by itself changes rapidly. It is a very unstable system. What we study doesn't always help us predict very much, but it helps us understand what is possible."

I pressed, "Is it possible for us to enter an ice age?"

José held in a smile, saying, "Sure, anything is possible."

He reached into a bag of trail mix on the table, thumbed peanuts into his mouth as he eyed me. Chaos man.

"How fast does rapid global change happen?" I asked.

"One or two decades is what I would call abrupt climate change," he said. "The earth by itself makes these jumps, even without us.

Right now we are tinkering to the point we could initiate a jump on our own."

"But we don't know where that jump would lead?" I said.

Koni over his notebook said, "*Ja,* exactly."

José: "A jump could go substantially higher or lower, both possibilities exist. There are some computer models that say global warming can lead to another ice age by disrupting climates."

Nico, looking up from his laptop: "But are they using Mac or PC?"

That made José laugh, practically snorting with delight. That was how much José trusted computer models to give a reliable prediction. He had already told me the only model he really trusted was the actual earth to which Koni had said even that is flawed and unstable. But, in their defense, José said he is also a fan of models, knowing they at least point toward probabilities, not just possibilities. They are the best guesses we have.

Koni waved off my question, saying, "If we've done anything, we've stopped the next glacial period from happening by warming the earth."

With excess atmospheric carbons keeping temperatures up in a time when they'd probably be going down, it could be as long as fifty thousand years before an ice age returns. Maybe then the cycles will begin again, ice growing larger from the poles and reflecting more sunlight into space, which cools the planet even further and deepens the hold of glaciation as over thousands of years wet, cold winters and summers of little melt grip the Northern Hemisphere. But few see this sort of thing anytime soon. Good news if you don't like ice. Bad news for all the predictable and unpredictable climatic breakdowns and reversals that could happen in order to keep things warm against any natural trends in the earth. One way or another, abrupt climate change appears to be at hand, what José refers to as a jump. Where do we land?

José would not outright discount the possibility of a new glacial period in our own time. Climate systems are too convoluted to tell exactly what they will do. Is a nearby ice age conceivable? Of course.

Probable? Not really. But José had to give a nod to uncertainty as a dominant factor. He said, "Science is not about common sense. It is about uncommon sense. It is about seeing what is not obvious."

In the middle of the conversation, Nico blurted, "In a hundred years we're fucked. I guarantee you."

"A hundred years?" asked Koni in his sage accent. "What do we know in a hundred years?"

"I'm just saying," Nico said.

The wind stopped during early morning. Everything fell still. After a third night of sleeping in what felt like the rigging and sails of a tall ship during a hurricane, I lay in my bag listening to nothing at all. No longer did a roar consume my every thought. It was a relief, but also a disconcertingly dead pause in what had been a very lively and unbroken conversation. Registering the sudden change, my nerves were involuntarily up.

A pocket watch was stuck in the snow beside my head on the floor-less bottom of my tent. A little before 6:00 a.m. I knew that much. I sat up as ice crystals flushed off the narrow, steep walls of my Scott tent, a canvas pyramid exactly like what Robert Scott used on his polar expeditions a century earlier. The sun was up, had been for hours, turning the inside of this sun-bleached once-yellow and now-orange tent into a rose-colored den. A soft mound of ice crystals had gathered on my sleeping bag, fallen from the breathing tube, an automotive hose stuck through a wall in case the storm buried me in here. Getting out of the bag was the hardest act, body freezing under one layer of long johns as I hit the air. I threw on layer after layer of icy clothing, forcing my legs into pants still stiff and clotted with chunks of pack snow. I jammed on boots, stomping them on hard-packed ground. Inner tent zipper, outer zipper, I stuck my head out to take a fresh, cold breath.

The air was filled with drifting crystals, almost a fog, but so thin the sky was clear and blue above. Ghostly breezes swirled into twirling shimmers of crystals. I shoveled out an exit and stood in the long shade of our tents, sun stretched across glass-plated snow. The storm had left a gallery of shapes behind the tents, teardrops and starships. I tried not to step on the more beautiful and elegant ones.

Our work shovels were all stabbed together like birthday candles between the holes we had been digging, our clumsy marks softened by the storm. We had most of the lost equipment shoveled out, snow machines hauled up from their pits. The two researchers from NASA had managed to land the day before, dropping through a hole in the storm. They brought with them a fresh snow machine that we unloaded from the plane and used with a come-along to pull out the others. Soon we would be able to get out to service and download data from remote stations.

In the kitchen tent, Koni was already at the table drinking coffee down to the dregs. José removed his hands from behind his head and offered a greeting as I came through the door. "It's our man of letters," he pronounced.

José could be reading off a cereal box, and I would enjoy hearing his voice.

Koni nodded with a firm "Morning."

"It's so quiet," I said.

As I stripped layers, Koni glanced at the Plexiglas window still coated in ice but not banging back and forth as it had been doing ceaselessly the last couple days. He said, "You think something is wrong, like the world stopped spinning."

"What causes this?"

"Could be the calm before the storm."

Koni said we might be in a low-pressure zone blocking a bigger, fiercer air front or, far more rare, wind patterns could be about to change as water-fat blizzards start blowing in from the west coast. Ice climates, especially in Greenland and especially during a swing season such as this, are particularly unstable, prone to rapid, unpredictable changes. As Koni had tried to claim on MSNBC, it can be naturally warmer here than the East Coast. He said sometimes it rains this time of year, moisture carried on a warm wind coming off the open ocean, and that, he said, is the worst of it: "People have to hoist your body off the snowmobile because they find you literally frozen in place from your ride back."

No spring rain would come today, though. A light ground wind was starting up from the west, cold, dry air flowing from higher on the ice sheet. Morning arrived slowly as the rest of the crew staggered into the tent. The boys from NASA were a friendly pair. They entered with folders and laptops to have with their oatmeal. They were both middle-aged, middle-American fathers of young kids and without any obvious accent, not so much unlike me. After breakfast, maps were laid out, and discussions were had about timelines and how long it was going to take for Koni to repair the snowmobiles so we could get out to the remote stations.

"Tomorrow we'll start," Koni said. "Today, we get out the last Ski-Doos."

They didn't need me for that. There were only a couple snow machines left, and we had dug them mostly out by now. I had been antsy to take a walk, have a look around the place at least a few steps past the perimeter of camp. This was the perfect day to look around.

I bundled up tight at the door and mentioned to Nestor, "I'm going to take a little walk."

There was a reason why I told Nestor. Koni would have suspiciously asked, *How far?*

"Where?" Nestor asked eagerly.

"Toward the sun," I said, waving my mitt slightly southeast as I opened the door and entered the silver-bright light of morning. A ground wind was flowing ankle-high, and I stepped through it as if it were a shallow wave gliding over a beach. I didn't have a plan. I just wanted to see a little more of the country. Maybe stand at the edge of camp for a while. It was too beautiful a day not to.

My boots pushed past the teetered, broken weather stations, their aluminum masts sticking up crookedly, wires exposed like the hair of some frightful giant beneath the snow. Beyond them, the world spread open. How often in your life can you look out and see nothing man-made to the very horizon, no road or tower or skein of fence? In fact, no mountain or valley. No stick or rock. It was a vacuum. I kept moving as if pulled into it. I stopped every several steps and glanced back. Nobody came out of the kitchen tent. I continued.

For half an hour I followed the sun. I became aware of it arcing through the sky. Conscious of my position, every ten or fifteen paces I paused to look back, aligning myself with camp so at least I could see how far I'd come. With a blazing golden light ahead, I could feel my place on the globe, my boot steps tromping across the bright curve of the Northern Hemisphere as it leaned toward the sun on a gently tilted axis. Wind blowing into my face was light but freezing, leaving long icicles off my mustache as if they were hanging from the eave of a house. After a while, the farther I walked, the less camp receded when I turned around, as if I were reaching some terminal number, space multiplied by distance equals infinity. I kept a keen eye on a few low clouds speeding past, knowing weather could turn fast. As newcomers, Nestor and I had been warned about abrupt changes by Koni, who said a whiteout can close in on you in a matter of minutes. There wasn't going to be a whiteout. I was betting on it. I felt I was on the safe side of being stupid.

Camp looked as if I could reach out and pick it up with my fingers. I lifted binoculars and could see two human figures bent over working, probably NASA out getting their GPS sensors in order. Koni no doubt had a bead on me long ago, and I imagined him looking out at me with his own binoculars, saying, "What the hell does he think he's doing?"

What did I think I was doing? Dr. Wegener, here I come.

The musical notations of my tracks extended across the plain of the horizon behind me, but they did not lead straight back to camp. Instead, they arced away as if I had walked here from somewhere to the northwest. It was the sun that did this. I had been walking the curved leg of an expanding spiral by following a point that was moving in a circle through the sky. If I kept this up all day, the sun would turn me around backward before it ducked under the horizon for a few hours and then climbed back up to resume this ironical march. Not stopping, I could make it maybe one night with what I had. Second night, I'd be dead, frozen. But it wasn't so much thinking about death that intrigued me as what the path of that death would look like. Think of the curlicue tracers of subatomic particles, the way they sometimes spin in on themselves before their energy pops and they wink out of existence. Following the sun across the ice sheet would look exactly like that. Shortening my stride, stumbling, I'd come to a winding halt.

I thought about stopping. I really did. But I had come so far already, nearing an hour, and the world was becoming larger with every step. Please, just a little farther. Being this detached from anything solid was a sensation so large and seamless I could only equate it with being an infant, learning at every moment how expansive and endless this world is: a tiny hand opening to realize *touch*, eyes detecting *distance*, ears hearing the voices of those parents who you had only heard through liquid. I saw the world so clearly, so fresh. Light clouds scuttled across the surface miles away, but my path remained clear.

There have been times when most of the world looked exactly like this. At least twice in the distant past (as in 600 million years ago) climate balances tipped so far from any equilibrium we are comfortable with that oceans almost entirely iced over and all land was covered over.* The cold would have sucked clouds from the air, leaving the planet bald and smooth as an eggshell. The so-called snowball-earth hypothesis came out in the 1990s and is now well on its way to becoming an established scientific theory. As you would expect, widespread extinctions accompanied these events. The bulk of extinctions appear to have occurred just before glaciations, suggesting it wasn't the ice itself that caused biotic turnover, but the change getting there. Again, it is the danger of the jump José had spoken of.

---

* Evidence of these hypothetical snowball-earth events comes mainly from glacial debris found in rock formations dating to that time and positioned in what were then tropical landmasses. These episodes would have likely been caused by an extreme tilt in the earth's axis, where the tropics may have actually been colder than the poles. Or the cause may have been runaway ice ages where positive feedback loops could have cooled the planet as more ice reflected more solar radiation into space, until the earth's surface gave up far more heat than it stored.

A deep boom shot out beneath me, and I stopped. I heard the course of the sound, an invisible crack issuing south to north. It seemed to cross the entire plain. Was that me? I suddenly became aware of being a fragile body in an overstuffed costume standing out on the ice. There weren't supposed to be any accessible crevasses around here, the snow too deep to break through. But what did I know about this place? For all I knew, some carnivorous species of hotheaded mole lived out here, traveling in packs, killers of the frozen waste.

This seemed like a good time to turn around. As I started back, another low crack split through the snow. It was a snow sound, not ice. I had heard similar sounds in avalanche conditions, a troubling boom under skis on a mountainside. But this was no mountain. What would be the analogue for an avalanche on a plain? Was I about to plunge through into a bottomless blue crack? Koni would later say these sounds happen all the time, plates of snow miles across shifting.

When I reached camp, light work was continuing down in the holes. When Nico saw me, he swatted the air with his hand and said in his characteristic French accent, "Spankey, spankey."

"Am I in trouble?"

The two from NASA were down in the hole looking up.

"What did you see out there?" one asked.

"Did it look any different?" asked the other.

Were they making fun of me?

Koni came out from the work tent and said, "Craig, can I speak with you over here?"

Behind the tent, his sunglasses looked at my sunglasses.

"Did you take a GPS with you?"

I said no.

"A radio?"

No.

He began chopping my shoulder with his glove. "I thought you were a wilderness person; did I misjudge?"

I didn't answer that; it really wasn't a question. Koni laid into a litany of reasons why what I just did was unwise, a rescue that would have to be mounted if I disappeared, danger to the crew, to the work that had to be done here. He did this politely where no one else could see us. I told him I was sorry and it wouldn't happen again. I wanted to tell him how beautiful it was and that I had been right about the weather—it had stayed clear the whole time—but I didn't dare.

Koni gave me a fatherly grip on that same shoulder and said I was a good worker and he would hate to have me die on him. I felt awful and relieved at the same time.

I retreated into the kitchen tent to hide my shame. José was just getting up from the table, closing his laptop. With the wind still down, he had heard everything outside.

"I feel like an idiot," I said.

"Mm-hmm," José agreed as he moved through the kitchen and suited up at the door. When he glanced over with his grave eyes, I was prepared for even my compatriot José, the chaos man, to accuse me of endangering the mission.

"Don't be too hard on yourself," he said. "When I first got here, I did the same thing."

He pulled on his gloves, cracked the door, and stepped into the sun.

Two days later, Koni threw me a GPS and said, "We need a hand at JAR 1."

JAR. Jakobshavn Ablation Region. A remote weather station. Koni had Frankensteined the snow machines back together, and gentlemen, start your engines. Lead scientists and their help shot out in all directions to service and download their remote sensors, hauling behind their snow machines sleds loaded with tools, ladders, and cases of electronics. From the back of a swift, snow-grabbing Ski-Doo, I clutched Koni. Looking around his sleek fur-lined leather cap and the strap of his ski goggles, I saw a larger country unfold. We raced miles past the meager distance I had walked the day before. Camp disappeared behind us in the three-mile-long swells of ice, oceanic rises and falls, where we half bogged down in powder drifts and launched airborne over wind curls. I hung on tight, trying not to grunt in Koni's ear.

The cause of these three-mile-wide waves in the ice sheet, the ones I first noticed when I landed at camp, was a mountain range thousands of feet below. As the ice sheet moved over those buried mountains at about a hundred yards a year, the ice surface became a series of standing waves, like a river flowing over an invisible boulder. The ice was right now moving at about one foot per day. We had been joking about the speed around the table one night, clutching whatever we could as if we were being swept away, a tent-full of laughing nerds. When the camp was first built in the early 1990s, it was at the highest point in the region and had a gentle vantage of everything around. Now, a decade and a half later, it was a full mile away from where it started, anchored into flowing ice where it surfed down the slope to the bottom of the wave. Koni said he'd be long retired by the time it rode up the next side of the wave and topped out again.

Ahead I saw a cloud hugging the ground like a whale. Only after a few minutes did I realize it wasn't a cloud, nor was it a whale. I kept waiting for us to pass through it, but instead we rose onto it. It was the ice sheet itself lifting, and we rode up its smooth slope. The farther east we traveled, the shorter the wavelength of these great ripples became, just like on a stream riffle where big waves become smaller, tighter, until they play out. We plowed up and over their long miles until reaching JAR 1, a metal mast decorated with sensors taking temperature, humidity, snow height, air speed, and GPS coordinates to record gradual ups, downs, and side-to-side movements in the ice. Only the mast had fallen over and was mostly buried. Reports had been coming back from almost every station like this where they were either melted out from a surprising warm spell last autumn or bent almost flat by the same winter storm that damaged camp. We had to dig them back up, chipping their structures out of ice, careful not to split wires with a sharp pickax. As I looked around for a spot to start digging, Koni waved his hand for me to get out of the way of two wind-speed indicators still sticking up from the snow, their blades spinning swiftly in the wind. The lee of my body was causing their tails to wobble, uncertain of the direction of the wind.

"They're recording right now?" I asked as I stepped back, fancying that the brief shape of my presence might be added into climate models, maybe even picked out with tweezers by somebody like José looking for the oddest aberrations.

"Yes, they're still recording," Koni said. "Don't get in front of them."

It was an all-day process putting this station back in order: constructing a steel-leg tripod, hoisting the thing out with pulleys and come-alongs, drilling a new and deeper anchor, and raising the ten-centimeter-wide metal pole upright again. The remainder of the afternoon, while Koni used his wire strippers clipping and splicing inside the control box, I marched around stabbing a probe in the snow, trying to find the tower and the box for a lost GPS station that toppled somewhere nearby. As I worked poking holes, hoping to find something hard, the clear sun rolled into a rainbow-colored circle of ice crystals. The spectacle took up a quarter of the sky, a halo floating over the ice sheet.

I once worked several seasons at a museum excavation, digging out the bottom of a cave floor down through a couple million years of bone deposits. We uncovered one massive ice event after the next, revealed by the fauna we found. Camels, cheetahs, and thousands of rodent bones tied to species connected with certain climatic conditions told us whether we had reached a warm phase in earth's history or a cold one. With a stratified wall exposed, counting up the different cold-weather and warm-weather species we had dug up, we identified seventeen distinct ice-age-like periods that prevailed over the earth in that time. These were not just the appearances and disappearances of whole ice ages but intermissions, endings, restarts, and interstadials.

These are what you get when the earth is in a glacial mood, as it has been for the entire Quaternary period, the last 2.5 million years.* A planet with ice on it tends to be naturally diverse, erratic

---

* In a key overview on climate changes from the Quaternary period starting now and going back 2.5 million years, Jonathan Adams, a paleobiologist from Rutgers University, wrote, "It appears that the climate system is more delicately balanced than had previously

even. Many, such as ourselves, learned flexibility in order to inhabit diverse niches and to survive major environmental changes. You might say modern humans are purely a product of this time in earth's history, that we were born of ice and what it brings, what it takes away.*

After we spent the day servicing remote weather stations, it was two in the morning before dinner was finished and most everyone had left for their sleeping tents. We were taking turns at dishes, and it was my night, sudsing bowls and wineglasses in a tub of warm, melted snow. Koni and I were the last ones up, coffee drained, wind howling outside. Wiping hands on a towel, last of the dishes put away, I sat with him at the round table, where we talked about his work. For most of the night since we got back, he had been writing a climate paper via computer with Jim Hansen in the United States, reporting on the science of how far is too far with CO_2 levels. He was trying to help determine what might tip the balance irretrievably into a new planetwide climate era, and what might not. As a lead author for the Intergovernmental Panel on Climate Change, Koni never seemed to be off the hook. He had heads of state and other high-ranking elected

---

been thought, linked by a cascade of powerful mechanisms that can amplify a small initial change into a much larger shift in temperature and aridity." Adams found that in at least the last 150,000 years there have been a few big, rapid climate changes that, if they happened now, would drastically change our world before our very eyes. How does this bode for our future? Adams wrote, "Even if one knew everything there was to know about past climate mechanisms, it is likely that we would still not be able to forecast such events confidently in the future."

* Nearly every species on the earth evolved from or adapted to conditions of glacial-interglacial swings. The Quaternary has seen extensive evolution caused by this climatic seesaw where some species stayed fixed in stable habitats, while most, like ourselves, moved long distances over time, pushed and pulled, creating a flourish of genetic interactions through brief adaptive radiation. Although the ice ages themselves pushed plants and animals out of large regions, causing spikes in extinctions, the rapid retreat of glaciers then opened expansive, suitable territories for repopulation through pioneering and reoccupation. On the other side of the argument, Quaternary changes in insects show little evidence for heightened evolutionary change. G. R. Coope of the University of London has written that there would have actually been "little imperative for any evolutionary change" in that time.

175

officials at his camp, world policy makers landing here to find out what needs to be done to save the earth, or at least their election campaigns.

"They think only in two years, election cycles," he complained. "What is two years?"

Then he smiled, remembering when he had danced here during a birthday party with one of the female graduate students. "It was to Abba," he said, laughing. "Right over there."

He thumbed over his shoulder to indicate exactly where the dance took place outside. He then had to stop and correct himself, aiming his thumb at a different spot, accounting for how far the ice has moved and changed aspect since then.

A couple days earlier, a plane had landed and out came a parka-laden crew including the crown prince of Holland, the head of the European Space Agency, and the head of the Norwegian Academy of Science. They convened with Koni in the tent for a breakdown of global climate issues. What he tells people who visit is not that the sky is falling but that we live in a world of falling skies and it is best not only to know your options but to make moves ensuring the worst does not happen. He is not a fan of points of no return. He thinks we have options, such as supplying clean-energy technology to rapidly growing nations, which he believes will be most responsible for future changes in atmospheric chemistry.

Sitting in the quasi dark between sunset and sunrise, Koni rattled off the usual dire climate news he and his colleagues had been uncovering, only in a more palatable way than Nico, who had said we'd be fucked in a hundred years. Koni used more science when he talked. I was hard-pressed to call him an alarmist. He seemed simply pragmatic, knowing what needed to be done.

"No, no, we don't want to panic," Koni said. "People never do well panicking. We don't make good decisions that way."

I hung dish towels over the propane heater wondering why so many scientists see themselves as guardians for the people. They do know something the rest of us don't, attentive to finer processes and sharply resolved details of the world. They have the numbers.

But they do not have everything. I said good night to Koni, leaving him with his papers and his waterproof notebook. I suited up at the door and walked out to amber light just before sunrise. Lenticular clouds were building in the north above the hidden sun. Their beveled edges showed the turn of high-level winds moving in opposite directions from ground winds flustering around me. The earth was sorting, processing, weather being born and swallowed over the polar ice cap. José, a man of precise and mathematical intellect, had said to me, "The earth computes." Everything I was seeing was part of that computation, variables thrown into the system and added together, creating a constantly changing world. How much change, how fast, and where it goes were the questions asked at this camp.

The clouds looked ominous, like dark-bellied UFOs the size of cities. Nico would later tell me these clouds weren't that uncommon, especially in the spring, when changes came fast, every day an average half a degree warmer, icicles starting to grow with drips at the end, slowly released from their former freeze-dried state. These lens-shaped clouds gathered in the north above a sun still below the horizon. The clouds were of the type *Altocumulus lenticularis,* a sign of moist, stable air blowing over the top of the ice sheet hundreds of miles from here. The wind had set up standing waves as it passed across the obstacle of the ice summit, much as you see in the wavelike terrain of the ice sheet, only on a much larger scale. Again, a stream over a boulder.

As I dug out a passage into my tent, knocking slabs of snow from steep canvas, the cloud stacks continued growing like mushroom heads pushing up through the sky. Rings mounted one upon the next into the troposphere, their copper-gilt edges reflecting across the ice sheet's empty arc below. I stood from my digging and watched as the world went to nimbus around me.

Maureen Raymo, a paleoclimatologist and research professor out of Columbia University's Lamont-Doherty Earth Observatory, told me,

"My feeling is that there is never going to be another ice age as long as there are humans on the planet." Raymo, whose influential early work revealed the slow rise of mountains leading to colder climates, said, "Right now we are actively changing the climate in a very uncontrolled way, but I'm fairly certain that by the end of this century we'll have developed the technology to titrate the climate, to basically control the level of CO_2 in the atmosphere. We'd have a thermostat."

She was not so much relying on technology to save us as she was projecting how a smart and manipulative species might operate in the future as it learns more about how this planet works. The technology, Raymo admits, does not exist right now, but that does not mean it won't.

"If in two or three thousand years the earth started cooling," driven by slow changes in the earth's orbit around the sun, "why would civilization let an ice sheet cover Europe?" she posed. "I'm assuming now civilization could last two or more thousand years, but why not? It's lasted six thousand. My personal view, and not very scientific, is that by the end of this century we'll learn how to control climate and therefore will prevent any further dramatic warming, and therefore prevent any further dramatic cooling. To me the only question is how bad does it get in the short term before the mental might of humankind is thrown at the problem, and the political will? Realistically, how bad could that get? I think a few meters of sea level change might be the worst that can happen, but that's devastating. Greenland melts, the West Antarctic ice sheet collapses. It's going to be horrible for civilization, horrible for low-lying population areas. If we saw that happening gradually, I think you'd see people spring into action and deal with the situation. But the problem is that the West Antarctic ice sheet and the Greenland ice sheet could have these tipping points where they get into a runaway mode where they become unstable and melt really quickly, and even if you quickly cooled the climate, you couldn't stop it."

Just the mention of us controlling the climate, not blindly poking at it as we are now, but manually steering it, sent a small shiver down my back. "How do you feel about us having that kind of long-term, actual strategic control of climate?" I asked.

"As a moral issue?" Raymo said. "To me it just seems like the inevitable outcome of the rise of higher beings that can control their destiny. Why do you dam a river? Why do you build a levy? It's the same thing magnified. Why would you let an ice sheet come down over Scandinavia and bury your village?"

Raymo was not resolved about this, not telling me how the future should unfold. It was just what she imagined. "I imagine humans will keep things pretty much like the Holocene. It's really nice, everybody likes the Holocene. If they were smart, they'd get their act together, say we can't let it be like the Eocene because we can't deal with the sea level rise, and let's just keep it at the Holocene for however long humans last."

We're a smart species. We figure things out. I believe Raymo is right that if we stay on this path of science and general interference with natural systems, we may learn to intentionally steer climates. She was speaking directly to my own desire to suspend this beautiful moment on earth, keeping it the way it is forever. But I had mixed feelings as she and I spoke of such a future. What does it mean to manufacture a planet to our liking, assuming we earned the skill to do so? Something sounded wrong about stopping ice ages by our own will, relieving the planet of its ticktock tendencies. It might look good on paper, but it sounds as if something would be broken, a wrench thrown into the gears of a swiftly spinning machine.

The word is "geoengineering," the deliberate tinkering with planetary forces to counteract unwanted environmental conditions. It could be one of the more dangerous futures for the planet, considering Murphy's Law. Bill Bishop, a Ph.D. chemist and former chair of the Advisory Committee for Geosciences of the National Science Foundation, told me, "It's an experiment we shouldn't try. The problem is that we will make mistakes. Big mistakes. That's just how science works. Only we're talking about the entire planet. There are too many unknowns, and the consequences are too big."

In an effort to turn atmospheric chemistry back into a more viable, preindustrial state, there have been serious discussions of seeding the stratosphere with hydrogen sulfide, which forms droplets of sulfuric acid, scattering sunlight back into space, bringing down global temperatures. Iron is being dumped into seas to stimulate phytoplankton growth, thus fixing carbons and sending them to the seafloor as phytoplankton dies, an effort to reduce mobile greenhouse gases. Bishop said, "Imagine if we shut off solar input by ten percent and we got it wrong, and it should have been five percent. You want to talk about an end of the world, there's one for you, a major geoengineering failure."

In Greenland a proposal is on the table to cover parts of the ice sheet with reflective material, bouncing back sunlight, blocking wind contact, and preventing melt. As a preliminary test, a team helicoptered in enough polypropylene blankets to cover about thirty-three thousand square feet of ice near Kangerlussuaq. It worked. The ice beneath the blankets did not melt.

At the round table at Koni's camp, we discussed this crude geoengineering project. José said, "It is the tactic of the dog. When it shits, it covers it up."

"It was the wrong message," Koni said. "Thank you, but no thank you. They think technology will solve everything, but there are other things we need to deal with, like consumption. They burned fifty thousand dollars just for the helicopter. The experiment was quite successful, reflecting solar radiation back off the ice cap. But melt is only fifty percent of the loss. The rest is movement and collapse."

José: "It gives people false hope that climate change has been fixed. I guarantee you there are millions of solutions. Here is a simple one: increase the efficiency of automobiles. If we go from twenty-five miles per gallon to fifty, we save eight million barrels a day. We now have to import ten million a day, so we cut foreign dependence by eighty percent, and we reduce emissions. That is one solution."

The stations felt lonely wherever we left them, as if we were dropping probes on another planet. Every day was a different world for us. Some days we traveled through light so diffuse you could hardly tell you were moving. Only the strike of snow-machine skis gave any sense of progress. Other days consisted of nothing but wind and the bright point of your own shadow beneath a round turquoise sky. Patchwork clouds quilted the ice, their islands of shade roving east to west.

I went out setting new GPS units with the NASA guys on a fog-crystalled day. It was all by the book, probing the snow first to make sure there were no crevasses below before anyone got off and started working. We had stopped inside what appeared to be an ice cloud. It was a dimensionless space with no up or down, no side to side. Wind sounded like ripping sheet metal, but somehow I had become accustomed to its constant barrage. I peeled myself out of a snowdrift that had built across my body on the gear sled I was riding, tied onto the back of a snow machine, and I stood in this blowing place.

My job here, besides digging and helping with tools, was to wander off with my journal in my mitts and stand somewhere looking outward the way you do when you are on a very small island and can't take your eye off the sea. In this case it was within the radius of snow-probe holes from where I could see only curtains of snow opening and closing. I had begun to expect that something would emerge from the murk, that I'd make out the rack of a reindeer or even a troop of lost sightseers. But I'd have to shake the thought from my head because, honestly, it was never going to happen. There was nothing out there.

That evening, the ice storm was gone, reduced to a swirling film of ground wind. We were back at camp. I was out digging fresh snow into a metal bucket to take back in and melt. I paused, looking up from my shovel to study the horizon. Skies were clear, the sun a blinding brass orb sitting above the snow. In this clear light, the horizon was perfectly evident, and nothing at all was on it. Nothing was ever on it. I suspected you would go mad if you looked at it expectantly for too long.

That's when it happened. When something appeared.

While José was marching back from his tent, I heard a twittering sound. Unmistakable. Bird. I recognized every note, a full-blown warble. I turned from the setting sun and scanned the emptiness of the gray southern sky. There I spotted its flutter.

"José, José!" I shouted, almost involuntarily. "There's a bird!"

José held up a glove to block the wind.

"See it?" I shouted, pointing at this small bird, flycatcher-sized, as it flew behind the tents.

José laughed when he spotted it. "Ha, I told you. Darwin was right."

In that glimpse I saw that it was a passerine, a perching bird. But there was nothing to perch on out here, and in fact nothing else living at all besides us. I didn't see it again.

José called it a "genetic dead-ender."

In the morning I looked for its body.

In a sugar-snow-filled windblast, I checked between gear boxes and around the foot of the tents, thinking I'd find a flustered lump of feathers. Night had been a deep freeze. Not that the days weren't the same, but these brief spells of semidarkness brought the worst weather and plummeting temperatures. If I could only find it, maybe I'd bury it somewhere, or I'd leave it alone in its death place. Anyway, I couldn't find it. But as I knelt to look between snow-machine treads, I heard its call again. The same clear-throated warble rang out. No sickly call, this was a live, vociferous animal, and it had lived out the night, a miracle. I stood to search the sky for it, but there was no sky to see. Wind was up again. Morning ice was blowing. I didn't hear the call again.

Maybe it had been a second bird off course, same perfect warble as the day before. If I knew birdcalls, I would tell you its name, but I didn't know birdcalls. It didn't seem likely that two of the same bird had found their way here one right after the next. What would José the chaos man say? Observe it thrice and it might mean something.

I wondered if it had been blown off course, or if it had crossed of its own volition, knowing somehow the ice was melting and Greenland's habitable country was widening by the year. It could have believed it would reach the other side of the ice sheet and only learned here that it could not. Too late.

I acknowledged that tonight the bird would likely die, especially in this stiff wind. Tomorrow I would not go looking for its poor, small, snow-pelted corpse.

It was 4:00 a.m. when I woke. The sky was bright. This was my last morning on the ice. A helicopter would land to trade me out for another crew member, and I'd be flown back to the fjord-cut coast of western Greenland. I figured this was my last chance to have a good look around, one more time trying to grasp the inscrutable landscape of the ice sheet.

First one up in the morning, I zipped out and stumbled into a clear day. I'd have the place to myself this early in the morning. No generator cranked up, no door opening or closing from our tiny island of Quonset tents. I was glad to see the sky clear, good weather for a helicopter ride. Wind hustled past, but it was benign at this point, a song rasping against the shell of my hood. I took a viewing position at the lee edge of camp out by the shitter, a V of plywood boards set up to protect a bare ass from the wind as we crapped in a hole in the snow. There I stood and did not move. I rooted myself in place. Over time, shadows shortened and swerved, the arcs of spirals as the sun slid higher in the sky. The pee pole behind me marked passing quarter hours like a sundial, its frozen, wind-cleared stump lemon colored from eleven days of urine. I was a sundial myself, mitts jammed into parka pockets, boots planted, my shadow turning around me as I watched the horizon, still expecting something to rise out of it, a thundering herd, a peeping head, anything.

I waited for nothing at all, watching high clouds skim the upper atmosphere, and I heard a warble. I had almost forgotten about the bird. The call came from off my left shoulder. I turned to see the bird lighting on one of the boards blocking our ice toilet.

"You," I whispered.

Oenanthe oenanthe, the northern wheatear. It fought its feathers back into place, body leaning sleekly into the wind. It was not supposed to be here, was not supposed to be alive.

A third chance encounter?

José would tell me about the odds being no longer random. This was the same bird. It was surviving the nights.

"Looks like you have another question to answer," José would say, grinning.

Damn scientists.

Was this bird a sign of evolution or death? Evolution: able to cross great distances under perilous conditions. Death: Wegener and Villumsen draining their reserves until they fell.

With double-trill wing beats, the bird launched into the wind and flew toward the tents, flit-sail, flit-sail. It first tried to land on my tent, which was the nearest, but it couldn't find a hold. There were no tent poles or any exterior features to grasp. Perhaps it had been startled from a sheltered roost in the shitter and was now trying desperately to get out of the wind. It left my tent and darted through back drafts between the other ones. Trying the peak of Koni's tent, it failed there, too.

I believed I was seeing a last scramble to survive, and any moment the wheatear would crumple to the ground, fluttering as it died in the snow. Against the forces of the earth, an individual life seemed fragile. I was worried for the bird, seeing a flight pattern that seemed to verge on panic as its wings knocked snow down from the steep walls of Koni's tent. Then it surprised me. It shot up in the air and spread its wings wide to catch the wind as if to say, *I will not die here, not this way, not with you watching.* With its small, angelic form, it made a scissor cut of bird shape across a wide arc of sky. With strong wings it peeled into higher winds, tacking across the ice sheet until it disappeared, leaving me here frozen still.

6

Species Vanish

How shall the heart be reconciled to its feast of losses?

—Stanley Kunitz

GRUNDY COUNTY, IOWA

ANGUS STOCKING, A gentlemanly urbanite from Madison, Wisconsin, found himself in the unfortunate position of having agreed to a short foot trek with me in central Iowa during a particularly warm July. He carried on his back enough gear for a few nights, and his eyeballs felt as if they were sweating. Dressed like a scavenger, bandanna covering most of his face, he marched through the interior of an industrial cornfield trying to think of being anywhere else but here. But there was nothing else to think of other than corn. Hot rows of veins and blades hemmed him in, parting around him with supple resistance. Plant whorls spun upward as if trying to strangle the sun, stalks grown eight feet tall closing in over his head. It was hard to breathe in the heat, his bandanna soaked with sweat and scurfy with corn pollen. He thought that perhaps he had never been more miserable in all his life.

Leaves as long as a person's arm draped across one another, and he pushed through them with his hat brim pulled down, leaves slithering around his sunglasses. The corn was one kind of green, a sun-shot parrot-feather lime magnifying daylight all the way to the ground. Not that you could see the ground.

The visually impenetrable mosaic of shadow and light might have been entertaining for Angus if not for feeling he might expire right here between the rows. He had counted on intense heat and humidity, anticipating temperatures that could top 110 degrees Fahrenheit at most. Instead, he had arrived during a meteorological anomaly

185

triggering heat warnings and advisories in thirty-six states. That particular week, nine hundred new record highs were set across the country, caused by a mass of high pressure trapping hot and humid air from the Gulf of Mexico. The nearest record breaker was forty miles away with a heat index of 129 degrees Fahrenheit.

Leaves clapped and rasped around him, closing in around the back of his neck as if dragging him in. Gasping behind his bandanna, he barged ahead, clasping his hands behind his back so they wouldn't get cut up. His dress was no affectation. After a few hours of being exposed, your skin starts to bleed, leaves so sharp they cut skin and cloth. In such a suspect environment as a pesticide-glazed genetically modified crop, you want to limit abrasions and open wounds. Rare as it is, these sorts of altered genes are known to pass horizontally into another species.

Angus briefly wondered why he didn't throttle me with his bare hands for doing this to him. This was my idea. All my other continent-hopping friends, climbers of Pakistani mountains, runners of Bhutanese rivers, laughed when I said I was looking for someone to join me backpacking through genetically modified cornfields where once-priceless soil now consists mostly of high-yield petroleum products. Not to mention there is the small chance of genetic transfer from the corn into your own genome.

Angus said yes.

Now he was wishing he hadn't.

I was ahead of him by a short distance, and one row over, laying a course through this six-hundred-acre field. Shoulder-width rows had been planted mechanically from a satellite-guided combine, little left to chance. This was the richest, most prime farmland in the state. When modern farmers first plowed Grundy County soils in the mid-nineteenth century, they said it sounded like firecrackers going off, all those prairie roots snapping as topsoil folded back like a blanket. Water is naturally occurring in the ground, no need for irrigation. It precipitates right out of the air in perfect daylight.

With a full pack, there was hardly room to turn around.

I stopped. My heart was beating hard trying to get the heat out of my core. I yanked down the damp bandanna so I could breathe. That didn't help. Everything was liquid, my flesh, my clothes, the plants around me. I felt as if I had just climbed out of a hot tub fully dressed, head spinning a little. I don't think I had ever been so hot and uncomfortable in all my life. A washed-out, half-white sky sparkled through the overlapping ceiling. If I set down my pack and stood on it, I could have stuck my arm up through the tasseled roof, a lone hand waving around in the sunlight.

I crouched, pinching the ground. It felt like dirty shoe polish, moist enough to hold a boot print. This was not the soil that farmers were pinching a century ago, nothing like it. This was now a created substance, an anthropogenic substrate. Beads of clear water formed along cornstalk seams and dripped down ribbed staffs as smooth as plastic. Each landed in the next leaf sheath down. I try to avoid humid geographies during hot seasons. I'd choose Greenland's ice sheet over this any day. But hothouse earth was having its say, harkening to other times in earth's history when humid forests stretched from pole to pole. Only this was very different. Other hothouse periods have been generally accompanied by high levels of bio-diversity. (The early Eocene, in which mammals rose to prominence, saw global forests expand from pole to pole, with redwoods and cypress swamps in the Arctic.) In this cornfield, I had come to a different kind of planetary evolution. I listened and heard nothing, no bird, no click of insect.

A thrashing sound finally came from behind, quiet at first but growing louder quickly. It was Angus. He sounded like an ox. He was breathing hard.

He stopped when he saw my face peeking through the next row.

"It's not really the deprivation chamber you'd expect it to be," he said as he pulled the bandanna and inhaled through his teeth. "No, it's much worse. It's like a sensory magnification chamber."

The meteorological anomaly Angus and I had walked into was being called a heat dome, a two-thousand-mile-wide cell of hot air sitting on North America like a bubble stuck to the bottom of a pan about to boil. It had shoved the jet stream up into Canada and was drawing hot, humid air up from the Gulf. Corn here liked the heat, a global-warming species, hybridized for five thousand years dating back to its humble wild grass beginnings in the Valley of Mexico (and now engineered so its genes contain those of an insect-killing worm as well as cellular resistance to herbicides). Farther south, it wasn't liking this heat. In Kansas, corn was burning up in the sun right now, leaves parched, yield shrinking by the day.* But central Iowa had so much water hanging in the air that drying up and burning was not a possibility. The hotter it became, the more moisture rose, turning this field into an atmospheric hydroponic.

Angus looked at the sky. Not being able to see it, he asked, "What day of the week is it?"

As if the position of the sun would have told him.

"Tuesday," I said.

"Tuesday? I could be dancing tango in an air-conditioned lounge tonight."

By trade, Angus was a skilled if not slightly obsessed land surveyor and a writer for the infrastructure industry—sewers, bridges, tunnels, and so on. He also ran a personal blog on existential paranormal conspiracy-theory phenomena—crop circles, UFOs, Illuminati, and such. I couldn't think of a better partner in a biotic wasteland.

Still pondering the sky, Angus licked sweat off his lips and combed his hand through his graying patch of a goatee above the loop of sweat hugging his neck. "What did you say to the Owens about us spending nights at their house instead of out here? When they said they'd come pick us up and fix us dinner and then drop us back out here in the morning? Oh yeah, you said no. You said we'd be fine."

"I didn't say we'd be fine," I said.

Mr. Owen was the farmer who gave us permission to backpack across his cornfields. He grew a combination of DuPont and Mon-

---

* In a 2012 article in the journal *Science*, a team of authors led by John Beddington, chief scientific adviser to the United Kingdom, wrote, "Global agriculture must produce more food to feed a growing population. Yet scientific assessments point to climate change as a growing threat to agricultural yields and food security." Since current climate change is widely believed to be driven by human causes, this creates a nonlinear feedback loop with agriculture, far from a fundamentally stable state.

santo stock, depending on which plot you're in. We were in DuPont corn now. It didn't look any different to me. When Angus and I promised we'd do our best not to damage the plants, Mr. Owen just laughed and said, "You can't hurt anything in there."

He was right. The leaves did not break without your leaning right on their nodes. They resisted, making travel like pushing through Jell-O.

"Two nights, three days," Angus said as he pulled up his bandanna, covering his face to the bridge of his nose, where he tucked it in under his sunglasses, no flesh showing. "We'll call it a long weekend."

That was a day less than we agreed.

He started moving again.

"That's three nights," I said after him, reminding him of our deal. "Four days."

But he was already gone, swallowed by corn.

We were walking a straight-lined labyrinth in a state more than 90 percent covered in agriculture. Whole landscapes were erased to make this, ecosystems tanked. This had historically been tallgrass prairie, one of the largest and most diverse biomasses in North America where a person on horseback could not be seen for the height of the grass. Timber surveys taken in the late nineteenth century found no trees here in Grundy County, only grass and marsh. After walking another half hour, I caught up to Angus, and we stopped. I shrugged off my pack and crouched with my head just above the first rung of leaves, where I could see a little farther, looking under the skirt of the corn. A host of memories returned, thirty million acres of tall grass, soil a tangle of bulbs and roots. Cordgrass, switchgrass, blue joint, side oats, brome. These weren't my memories. They were of a place that existed before botanists pressed the last specimens into their books.

Angus dropped in the next row over and rolled to his back on the damp, pliable ground, letting out a groan as he unbuttoned his shirt.

"This sucks," he said.

He had driven down from Madison to join me on this mission. It hadn't been half a day yet, and he was ready to quit. He'd probably be able to find a cold beer within an hour. . . . *I could walk to a highway, hitch to the nearest town, and whatever they've got on tap in a cold glass mug would make all this go away.* But he had a bargain to keep.

The only place where you could find breathing room was in the foot of airspace between the ground and the bottom leaves where Angus had taken up residence. Shirts opened, bellies white, we lay between rows breathing.

"Somehow this is not as fascinating as you might think," Angus said.

I didn't reply. He seemed a little on edge.

Weeping with sweat, I rolled onto my side and wobbled my eyes through this one-foot-tall landscape of green broom handles evenly spaced (a mathematic pattern no matter which way you looked). It was really the only view you could get, a horizontal plane of weird, protuberant brace roots growing out of lower nodes of stalks. The corn appeared as if it might pull itself up and start wandering around like urchins. Then it would be chaos. Without rows, you might never find your way out of here.

I rose to my hands and knees, my skin already coated with a darkening rind of soil, not the most healthy of substances. I felt like a grimy amphibian, and I crawled between stalks, crossing to the next row.

"Going somewhere?"

"Looking for wildlife," I said. My object here was to see what else, if anything, was alive in these fields.

Fourth row past Angus, I found a mushroom the size of an apple seed. With elbows in the grunge, I crawled closer. It had a dreary-colored gnome hat, the only thing not corn. It was a waxcap, not a kind of mushroom I'd expect, a family used as an indicator of pollution, highly sensitive to common agricultural chemicals. Judging by the fact that almost nothing else was growing in here, I couldn't imagine this being healthy waxcap soil. But here it was, a survivor, a pioneer even.

Tiny mushroom, family *Hygrophoraceae*, check.

What else lived here? I lifted up leaves, scanning the undersides near the stalks, and after several tries I disturbed a small cobweb spider with a pearl of an abdomen hanging from its thimble-sized web. I caught it sucking juice from a captured crane fly. It quickly retreated into the seam of the leaf, which I lowered, giving the spider back its meal.

Cobweb spider, family *Theridiidae*; crane fly, family *Tipulidae*, check.

These fields are managed for maximum yield, and anything but corn (or soybeans in adjacent plots) is considered rogue and either destroyed or prevented from growing or surviving. There were originally three hundred species of plants in tall grass, sixty mammals, three hundred birds. Kinds of insects numbered over a thousand. Nosing along the ground, I found one busy red mite the size of a dust mote hurrying across barren earth, up and over shredded pieces of old corncob. It felt like another planet entirely, a bizarre world of little nomads at the base of monumental corn stalks. There was an ant, too. It was so small you couldn't pin it to a specimen board. Industrial agriculture miniaturizes species, allowing in only what is very small or what is uniquely adapted.

Early afternoon we walked into an agricultural drainage, a swath carved from the middle of the field. Pheasant hunters call these things *sloughs*, gaps in the corn where you can get a shot off rather than in the tangle of standing corn. It was a dry river of waist-deep grass winding north to south, the first open breath under the sky. Among nodding foxtail, clouds of grasshoppers sprang ahead of us, hundreds of them clicking to the air with our every step. The open sky was strange, prismatic, as if sunlight were passed through white glass. There were no bees—notoriously missing from these agricultural galleries—but I did see a delta-winged katydid clutched to a grass stem and a black stink beetle trundling along the ground. The vast majority of living, moving things were grasshoppers, and they were everywhere in this open passage. I'd never seen so many, as if this were their refuge.

You don't see grasshoppers back inside the corn. Pesticides keep them out. But I could see along the edges of the drainage grasshoppers were trying get in. They had chewed open tips and were entering fresh ears, their mechanical mouth parts sounding like tiny scrapers chiseling at milk-hard kernels. Toxins applied both from crop dusters and what's been added to the genes of corn itself would ultimately kill most of them, or at least drive them back, but some would survive. I was watching evolution in progress, sped up as if this were a petri dish stewing the genetic future of our earth.

The first identified resistance to crops engineered with the *Bacillus thuringiensis* pesticide came in 2009 when a pink bowl worm in India developed genetic tolerance and spread rapidly. Large agricultural regions in India found themselves unable to fend off this particular pest. That is one risk of mono-cropping; the race is constantly on to stay one step ahead of the natural evolution of pests.

These grasshoppers had not achieved locust capacity, flying en masse across entire regions and consuming everything in their path. Every grassy tiling in Iowa looks like this in the summer, isolated genetic islands, unique pieces of biogeography where insects were in a race with genetic engineers for the chance to spread. Like the pink bowl worm in India breaking through a wall of genetic defense, life cannot be contained. But it can be killed.

I chose Iowa for a mass-extinction analogue because it is the most thorough picture of genetic exhaustion, the many organs of what was once tallgrass prairie removed and replaced with this. Extinctions are happening all the time, genetic lines terminated by running out of safe places to live, or climate collapses, or simply turning into a new species as the old one dies out over tens, thousands, or millions of years. The end-Permian extinction saw 90 percent of life purged from the fossil record in seas, more than 70 percent on land; its ending appears to have peaked over a twenty-thousand-year period. What picture comes to mind of such a termination? Is it windswept desolation, perhaps the dune desert I was walking through in Mexico, hangman trees standing like rotten crucifixes?

It would actually look like this, a lot of one thing and not much of any other. You might not even notice. There's always been life on earth, rampant even in the worst of times. Sometimes it is diverse, rich, and capable. Sometimes not, prone to catastrophic, trophic failures like the ones being seen in world oceans today.

In the last forty years, plankton counts have dropped to half their previous levels. Considering that plankton is one of the primary oxygen producers on this planet, this kind of loss has far-reaching ramifications if it fails to rebound.

The end-Permian extinction created its own form of monoculture with short-limbed, pig-sized lystrosaurs. These barrel-chested animals had large lungs, allowing them to thrive in a high-CO_2, low-oxygen environment. They make up sometimes 95 percent of remains found in fossil beds from that time, making it appear that the world consisted mostly of grunting, burrow-digging lystrosaurs.

Technically, extinction is the process of a species or larger familial group ceasing to exist. To become extinct is to have your unique course of genetic evolution cut off from wherever it branched away from its last distinguishable predecessor, which can go back tens of thousands if not millions of years. Dinosaurs were reduced to birds around 65 million years ago.* The American cheetah 11,000 years ago pinched out, after having existed for 1.8 million years, its closest living relatives now mountain lions as well as whatever cheetahs migrated across the Bering Land Bridge and made it to Africa. What of humans? If we disappeared, it would be a long way back to our nearest living relative to keep a resemblance of our line going. So many of our previous hominid links disappeared themselves as if methodically cut out of the record. A creationist's worst nightmare, our nearest surviving relative would be chimpanzees or gorillas, the closest living genetic connections we have.

---

* Mammalian clades made it through the dinosaur extinction sixty-five million years ago relatively unscathed. Though we tend to think it was mammals that fared best after that, birds actually made the biggest surge, the forms that made it through derived from small, bipedal, feathered dinosaurs like *Archaeopteryx* and *Deinonychus*. These dinosaur descendents have now become one of the most successful and diverse classes of vertebrates in the world, represented by 9,000 living species (mammals currently make up around 5,500 species).

Mass extinctions are often relatively slow, taking centuries to thousands of years to be felt, and in the case of the end-Permian die-off, another eight to ten million years was required to bring the earth back to previous levels of genetic diversity. They begin with what we were seeing in Iowa, with extirpation, not the total disappearance of species per se, but their being displaced, bottlenecked. Most organisms that once lived here still have a genetic presence somewhere in the world, specimens in greenhouses or grown on refugia, or seeds locked up in frozen compartments for an unknowable future. They may not become extinct immediately, but being pushed out of decaying or destroyed habitats eventually takes its toll. The concept is known as extinction debt, the delay between the stress on species and the final dwindling of the last survivors until the organisms disappear and are never seen again.

Edward O. Wilson, premier sociobiologist, wrote, "The one process now going on that will take millions of years to correct is the loss of genetic and species diversity by the destruction of natural habitats. This is the folly our descendants are least likely to forgive us."

"What time is it?"

"Two fifteen," Angus said.

"That's all?" I asked.

"Seven hours to sunset, my friend."

I sat up, looking down-row past my feet until the corridor quickly became too crowded with leaves and I couldn't see any farther.

"Crab spider," Angus said, reaching up a finger and poking at a leaf above his head where he could see its small shadow through the green.

"Thanks," I said, reaching for my journal.

Family, *Thomisidae.*

An almost imperceptible breeze moved across the top of the corn. Leaves gently sawed against each other. It was no relief.

The corn ears were not ready, too young and starchy for harvest.

I had tried one earlier, and it tasted like half-sweet sawdust, its milk getting in my beard, making me all the more nappy. This was seed corn. Mr. Owen plants this plot for Pioneer, owned by DuPont (pronounced "dew-pont" around here, accent on "dew"). Pioneer gives him the seeds and then takes the harvest, which it genetically tweaks for the next generations, trying to keep ahead of pest adaptations. Pioneer then sells the next season's seeds to Mr. Owen, who plants them in a different plot, where he grows his corn for syrups, fuel, disposable dining ware, bath powders, laundry starches, adhesives, and so on. One ear of corn does it all. You can even buy it in the grocery store, butter it up, and have it for supper.

I had asked Mr. Owen if his family eats their own corn, and he gave me his polite Midwestern laugh. To him, this was unabashedly *product*, not food. He might as well have been selling bumper crops of nails. But he believed in his crop, saying, "The world needs food." If this scale of production suddenly collapsed, so would human populations.

His wife, Mrs. Owen, had been married into this industrial agriculture family for forty years. A generous and subtly cunning woman, she checked out books from the local library to help me identify plants. When I asked her why she thought the state had recently changed policy from cutting and applying herbicides to highway right-of-ways and had begun fostering growth of flowers and native grasses, she said, "It's for the good earth."

As if I needed no other answer.

Her phrase stuck with me, *good earth*, a justification in itself, a reason for doing something.

After dinner one evening, Mrs. Owen took me aside in the kitchen when her husband was in the other room. Wiping her hands on a towel, she said, "I've been in this kind of farming most of my life. I don't think what we're doing is good for the earth."

Why would she say this? Because of cancer clusters and the decline of biologic diversity, because of the houses she has seen torn down, some on her own land, farm families bought out and their grounds turned into new rows to up the yield. Because this land is planted fencerow to fencerow, taking down old oak trees and sloughs, almost every last piece of ground is consumed. For all the food this produces in a hungry world, she was saying that payback is a bitch.

Atrazine is one of the most commonly used herbicides on a field such as this and has been found to cause reproductive dysfunction in amphibians, fish, reptiles, and mammals. It is the most commonly detected pesticide contaminant of groundwater and surface water in the United States. In 2001, twenty-two international scientists published their review of atrazine from North and South America, Europe, and Japan in *The Journal of Steroid Biochemistry and Molecular Biology*. They show a common association between exposure to the herbicide and the "feminization" of male gonads in many animals, and sometimes complete sexual reversal, where males actually become females. In a 2010 study reported in the *Proceedings of the National Academy of Sciences,* Tyrone Hayes, a professor of integrative biology at the University of California, Berkeley, and lead author of the review, showed that sexual reversal in frogs was directly associated with atrazine exposure. When the study came out, Hayes said, "And this is not at extremely high concentrations. These are at concentrations that are found in the environment."

More than seventy-five million pounds of atrazine are applied to corn and other crops in the United States every year. At the same time, cancer clusters are found throughout agricultural processing regions. Mainstream science has been reluctant to make a direct link between human cancers and herbicides and nitrates, most often reporting "potential associations."

Meanwhile, biodiversity, one of the key indicators for environmental health, has tanked in these agricultural regions. Biotic depletion doesn't end in the fields themselves. Fertilizers that are spread on these crops wash downstream in rains, flowing into creeks— pronounced *cricks* here, rather than *creeks*—and then rivers, eventually reaching the Gulf of Mexico, where they trigger the largest marine algal bloom in the world. That bloom, in turn, pulls oxygen from the water, creating a dead zone extending thousands of square miles, an extinction debt racking up on a scale you'd see only in the greatest environmental catastrophes.

While these fields are reaching out and killing the Gulf, they are also doing a job on monarch butterflies across their entire North American range as far as southern Mexico. Glyphosate herbicides like Roundup have seen increasing use in the Midwest over the past decade, up to 100,000 tons applied annually in the United States. The toxic molecule is absorbed through leaves where it inhibits enzyme production and kills any plant it touches. Crops have been genetically engineered to resist this herbicide (the gene sequence first patented by Monsanto), so in many places glyphosate can be freely applied. Between 1999 and 2010, milkweed, a seemingly unwanted plant, has diminished in the Midwest by 58 percent due to these herbicides. Milkweed is also the one plant on which monarch butterflies lay eggs, and while the host plant has declined, butterfly populations have seen a decade-long plummet, as much as 30 percent reduction in one year.

Being at the source of this kind of bad news, I was looking for what if anything was surviving, a thumb-sized toad, tracks of a whitetail deer. These weren't much, but at least more than nothing.

This is what every mass extinction in earth's history has looked like, cascading, wide-scale devastation but with signs of survivors, the beginning of the next world hanging on. Without survivors, there is nothing. Every organism that makes it through becomes the possible biologic foundation for what is to come. In the case of the end-Permian extinction, the piggy, tusk-toothed lystrosaurs are ancestors of mammals. Humans are here now only because this animal somehow persisted through the greatest extinction in earth's history. You could think the same with creatures now endangered. Those that don't survive leave a hole in future possibilities, biotic recovery or diversity kicked down a notch.

Most scientists agree that the earth is experiencing such a bottleneck once again. We are either swiftly approaching or well into the sixth mass extinction in earth's history. The pace of loss is estimated somewhere between 18,000 and 140,000 species per year.* Whatever perishes now is out of the running, while what lives lays the groundwork for the future earth. We are deciding what makes it through.

As I rested with the back of my head against the mashed ground, a brief shadow flitted past.

"Bird," Angus said from the row next to me.

"Bird," I confirmed.

We set our camp back in the drainage we'd found, dropping gear where soil beneath us held skeletons of PVC pipe buried to carry water out and keep the corn roots from rotting in damp soil. We put up our two tents on bare ground balded by an excess of herbicides. I stirred mac and cheese over a small can stove. Angus unscrewed a half-pint Ball jar containing a chestnut-colored single-malt scotch he'd brought along. He liked his comforts, Thai massages and fine meals. He leaned back on the ground and sipped, watching the first lightning bugs appear as dusk deepened around us.

"I used to be sure the world was going to end," Angus said. "I don't think so anymore. I think it's always been like this; it's always felt tenuous. Every generation wants to be the one; I spent seventeen years of my life knocking on doors trying to get people to accept my religion. I believed I was helping them to survive the apocalypse. I was sure it was coming."

Angus had been a Jehovah's Witness for those seventeen years, had a wife and kids, all church members. Single now, kids grown, and the church long behind him, he said, "I was preaching a religion that does not bring good into people's lives."

While he talked, a hummingbird-sized sphinx moth with a long proboscis fluttered at my left ear. Thinking me a white and blooming

---

* High estimates of 140,000 species lost per year are based on species-area projections, which have been criticized as possibly being a 160 percent overestimate.

night flower, it uncurled its tongue into my ear canal, a quick tickle, and it was gone.

"I think it's more tenuous than we imagine," I said, swizzling a finger inside my ear to scratch the itch.

"You think the world is ending?" he asked.

"Something big is happening."

"Something big is always happening."

"Look at our populations, though, our rate of consumption. Look at this corn and all that was lost here."

"We always want to be the ones," Angus reiterated. "We want our place in history to be the marker."

I said, "It might be true this time."

Reclining his head into his hands, looking at the sky, he said, "Maybe the earth needs a season of corn just like we need a night's sleep."

For a moment I considered his comment. Could Angus be right? Does the earth need the occasional downtime of a mass extinction? I was aware of the patterns coursing through time, genetic diversity increasing in a jagged curve for sometimes hundreds of millions of years, then bang, mass extinction and back to scratch. After that, the void fills rapidly with a world reborn from adaptive radiation, a whole new set of eyes and breathing stomata unfurled across the earth based on the few that made it through. It was how placental mammals got their toehold in the first place, bursting from the dinosaur extinction and leading directly to us. It was done through loss, a process of destruction and creation intertwined.

"I don't know," I said.

"My point exactly," Angus said, as if it were a harmless fact. "The only thing I can do to increase good in the world is create satisfaction and joy in what is happening around me. That's where I hang my hat after years of trying to do good in the world. This is the only way I've found that works. Not only is the world not my problem, it is not a problem at all."

He sipped his scotch and lightly smacked his lips.

His idea of good and mine were slightly different. Mine was a diverse, thriving world of many species, their genetic lines fanning back through time and ahead just the same. His was good scotch. They both have their place.

I visited with Edward O. Wilson at his office inside Harvard's Museum of Comparative Zoology, and I posed to him what Angus had said about the earth needing times of corn, times of extinction. Wilson said, "There is no such thing as rejuvenating a place by eliminating biodiversity, then letting biodiversity flow back in. We don't have the millions of years that would take, it's as simple as that."

Eighty-two years old, Wilson's voice was aged and patrician with a native southeastern accent. He said that Angus's point was a common question. Wilson said, "Isn't extinction a natural process anyway? Isn't it true that ninety-nine percent of all species that lived have gone extinct and they're replaced and so on? That comes from a lack of understanding of timescale. Earth before man had an extinction rate estimated at one species per million per year, and also the same rate of new species production, so there was this turnover. Spun out over three and a half billion years, sure, ninety-nine percent of the species have gone extinct, just as the vast majority of humans to have ever lived died. But these species left behind rather healthy, huge populations of descendants."

In other words, when species go extinct under anthropogenic influences, they are not leaving behind healthy, huge populations of descendants. They are leaving behind very little. Thus, mass extinction.

"You know, this notion that we're just another agent of extinction, people are doing their part extinguishing species like great storms or droughts?" Wilson mused. "We're moving from a thousand to ten thousand times the rate of extinction removal than before we came along. Another notion is that there are species that have just grown old, and like old folks you've got to let them go; you can't be spending large amounts of money trying to keep them alive, so pull the plug, let them die gracefully. That's a fallacious analogy made between a person and a species. All species consist of young, vigorous, healthy individuals. You go find a species right on the brink of extinction, and you're always astonished. They're young, healthy, vigorous. It's just that they're not being replaced because there's no place for them to breed, or because they're being shot out, or because there's not enough food for the young, or there's a bad new predator introduced. But when you look at the individuals themselves, what the species really is, they are just as ready to breed and spread as a species that covers half the continent. There's no difference."

Dinner was finished, pan on the ground where Angus and I had eaten only half of what I fixed. The air still felt like a sauna, no good for the appetite. The heat wasn't going away. Gentle breezes rasped through the corn. They were not comforting, still hot as day. Blue-green sparks danced up from the foxtail, and by 9:00 p.m. fireflies were all around us. Crepuscular bioluminescence, what would this world dare invent?

For want of something to do other than sit here dripping on ourselves talking about the world ending or not, we walked north along the tiling until we hit a boundary with a low-cropped soybean field. The dusk sky was huge on the other side of the fence. Tobacco-green rows rolled out like a bannered flag.

Big spade leaves brushed around our knees as we walked. The last of twilight faded. Countless lights of fireflies zinged upward like sparks and fell back dark to the leaves. The list of other organisms we spotted in the corn was short so far, fifteen altogether: sphinx moth, grasshoppers and their kin, spiders, black beetle, mushrooms, a thumb-sized toad we found in the dry drainage, deer tracks between rows, and whatever that bird was that flitted through the corn shadows earlier in the day. Now our world was dominated by these lightning bugs.

We stopped at a gentle rise. We made statues of ourselves, silent in the layers of heat and the indigo streak of the horizon. I glanced over at Angus. His form was as black as a nightshade against a gently quilted landscape where I could see one darkening horizon after the next. There were so many fireflies they looked like a star field in a crisp

201

night sky, miles of them laid out across soybeans until they blended into a scintillating electric-blue haze as far out as we could see.

"Just when you thought the world was ordinary," Angus said.

In the cricket-filled dark of this moonless night, fireflies had all but gone out. I was walking through a chamber of sawing insect legs, heading for my tent, when I was stopped by a loud snort coming from down the tiling. I couldn't see a thing. Again the snort. I heard hooves prancing. Whitetail deer. I had been seeing their slender, heart-shaped hoofprints. Family, *Cervidae*. This drainage, I thought, must be a corridor for night travel, shuttling from one field to the next. Our camp was in the way.

The deer paced in the blackness, curious and not sure whether it should turn and run or stay. Another snort. I took a step closer. The deer jumped away, and I listened to its pogo hop receding through foxtail.

The nearest piece of virgin prairie, ground left historically untouched, is around an old pioneer cemetery in Butler County, one county over from Grundy. The town that was there—Butler Center—is gone now, plowed under and replaced with crops, while its cemetery remains atop a subtle rise in the topography. Half the cemetery grounds have been left empty, let go wild, not plowed or burned in centuries, no graves ever laid there, just as is. It is a piece of original Iowa.

Ruth Haan, a woman in her eighties, was one of the last on a board of volunteers overseeing this plot. Locals warned me Ruth was losing her mental faculties. Her niece who drove her to the cemetery plot on a blistering summer day said right in front of her that she was getting a little loopy.

"Oh, honey, I just need a little help now and then," Ruth said in her sundress, the fat of her arms hanging like handbags.

She pushed her own wheeled walker across bumpy, mowed cemetery grass to reach the fence marking a refuge she'd known her whole life. "Volunteers haven't met for quite a while because most of us have died," she said. "I sure hope someone will keep an eye on this after I'm gone; it's the last piece around."

On the other side of an old tangle-wire fence stood a waist-high briar of shrubs, grasses, and flowers. It was not purely native by any stretch, but it was virgin, unplowed. Where invasives are weeded out by hand, native tall grass tends to be far more majestic, amber waves of grain that outdo any wheat field—their summer shadows flittered with birds. This cemetery plot, known to aging volunteers as Clay Hills Preserve, looked more like a wild and forgotten lot, a relief to my eyes among sheer horizons of monoculture. There were even milkweeds growing around the edges, favorites of monarch butterflies.

Ruth wore a yellowed straw sun hat, and I could see her wearing that thing for years, on her knees out here in the sun pulling up weeds. She stopped at the fence line and said, "Smells good, doesn't it?"

It didn't smell like corn. It smelled hot and writhing and of many forms of both decay and growth. The fence was half pushed over in places. Ruth guided her walker along it, surveying this multihued green so thick you couldn't see a sign of ground. With a liver-spotted hand, she pointed at a plume of fine grass and said, "When you find dropseed, you know it's virgin; they didn't plow it up. Wild parsnip, not native, shouldn't be in there. Some of these shrubs and yucky stuff—that grass over there is invasive. It's not supposed to be here and, oh, nettles, nettles, that's a shame. But, oh, look there, rattlesnake master."

She was pointing at a plant bearing prickly greenish-white balls of flowers. Pioneers believed its roots could be used to cure rattlesnake bites, back when there were rattlesnakes (though still plentiful in northeast Iowa).

"It's a wonder it's still here," Ruth said. "That makes me feel good to see that."

I helped Ruth along, holding her hand as she stepped from the walker and leaned against the fence, grabbing wire as she looked over at the rattlesnake master. "I wish it was prettier for you," she said. "It's pretty for me."

I promised her that to me it was beautiful. Forever-summer green of corn stood all around, a monolithic landscape surrounding this one postage stamp of a former world. It existed for its own sake, offering a different kind of nourishment than corn. I felt grateful to Ruth and her inheritance still here in the ground. I asked her repeatedly why she cared for this at all, why in a world driven seemingly by nothing but yield she or anyone else would want to save a tiny plot of virgin ground. Ruth found my question to be nonsense.

"Go on in and have a look," she said, as if that would answer me.

I stepped back to see where the gate was.

"Just hop over."

On the other side I landed in chest-deep grasses, finding a path where deer must come through. Even that was overgrown, hard to put down even a step, my hands brushing through a plume of a grass called prairie smoke—at least that's what Ruth called it as I waved my hand through its ghostly seeds. I waded into the middle of the plot, crouched, and sank my hand toward the ground, which was not really ground but layers of fallen grass in stages of decomposition. Down on a knee looking through the tangle, I thought this seemed more like life on earth.

What is the genetic value of a plot like this? Not much. A single grain of sand. But it swayed me. It was something different, a piece of memory rather than complete amnesia. I saw exactly what I believed would be here. Hope. Current depletion rates and the rapidly building insolvency of extinction debt now stand out as the most profound change the earth has experienced since the extinction of dinosaurs. Iowa represents the epitome of human impact, the world completely changed on a landscape scale, whole biotic geographies erased to feed our growth. Without the sentiment of refuges, all could be lost. The reason for this place was what Mrs. Owen had meant when she said *for the good earth*.

When I hopped back over the fence, Ruth was waiting for me

between the cemetery and this piece of tangled prairie. I helped her turn around to go back the way we came, back to her niece waiting under the shade of an apple tree next to her silver minivan, its door slid open to usher the old woman home.

Big bluestem and Indian grass, these tall perennial grasses that grew here once had roots twelve feet deep, one of the best terrestrial carbon sinkers on earth, more efficient at scrubbing the atmosphere of CO_2 than rain forest. There used to be bison, and the dominant central Iowa herbivore was, surprisingly, elk, as if before the nineteenth century this had been a refuge for the last and smallest of the mega fauna from the Ice Age. You can imagine the strong bones of elk antlers brushing through high prairie, nothing to see but their wearied tips above the grass heads. Bison and elk are now gone. And almost everything else.

By 1876, Iowa prairie was the exception rather than the rule. As efficiency increased, rows narrowed from wide enough to allow a horse and wagon to wide enough for a horse to now shoulder-width so a man (or deer) can walk through. Genetic tinkering has allowed corn plants to snug nearly against each other, leaf patterns maximizing the allowance of sunlight. New strains of straight-up leaves are beginning to spread, potentially allowing corn plants to be planted even closer, fields impenetrable, raw masses of life allowing few others to exist.

As long as the supply of water, sunlight, and fertilizer doesn't end, this goes on forever. But everything ends. Every inch of Grundy County soil, traditionally the benchmark for the highest price farm land in the state, took a thousand years of prairie ecology to create. Now it is being used up, modern yield impossible without massive application of fertilizers. Like almost every industrial farmer in the world, Mr. Owen uses anhydrous ammonia on the soil, a petroleum product that promises a high-enough yield to support the global food trade. It recharges nitrogen in the soil and is the largest chemical addition made to croplands. A critical ingredient in making methamphetamine drugs, anhydrous ammonia is so poisonous that a single breath of it can kill.

In the company of a state conservation officer, I asked Mr. Owen how long Grundy County soil could hold up. Both Mr. Owen and the conservation officer broke into slightly cagey laughter, glancing at each other as if I had just tripped over a skeleton in a closet.

Mr. Owen patted me on the back, "Oh, you don't want to know that."

He was right. I didn't. I wanted to go on living happily ignorant, unaware of how global agriculture could collapse Wall Street–style. If you want to see a model for human extinction, or at least a serious, rapid drawdown in global population, there's your window to look through. We don't need contagion or an ice age or even cancer clusters around agricultural production regions to do us in. An agricultural failure would do. A volcanic winter, a pandemic of agricultural diseases or pests that finally outsmart genetic engineers, or a simple failure of soil fertility. The lives of many are balanced on these endless green rows.

All things considered, we are one damn cunning tribe of animal, whose chances of survival I'd put as surprisingly high. I don't anticipate our extinction any time soon regardless of potential collapses. Besides microbes, few living organisms can spread so far and subsist on such a wide variety of consumables. We are made to survive. My concern is what we might do to the rest of the world in that endeavor.

I exchanged notes with Lisa Stein, an exobiologist out of the University of Alberta who studies the effect of microbes on greenhouse gases. I wrote to her, saying I was looking for a sign of hope.

Stein wrote back, "I realized while teaching environmental sciences in 2006 that we had reached the point of no return with global climate change. After a bit of panic and depression, the only other option was to accept what was happening and try to move forward.

It's interesting to be in the nitrogen field because it's the most obviously human-impacted biogeochemical cycle and we're able to measure the microbes reaching for a new state of equilibrium. So as a scientist, it's exciting to stand on the precipice between the Holocene and the Anthropocene. We're the only humans to have experienced a shift in geological epoch."

It wasn't the good news I had hoped for. Stein's mention of a shift in geological epoch is a somewhat new thought among scientists. It is a marker for the end of the Holocene and the beginning of a new phase for the earth, a recognition that humans have become one of the largest agents of global change. In 2008, a proposal was presented to the Stratigraphy Commission of the Geological Society of London to make the Anthropocene a formal unit of geologic time, which would officially end the ten-thousand-year-long Holocene. Some believe our new age began with the first atomic bombs leaving a layer of radiation around the globe, as if by the first strike of an asteroid. Others take it back thousands of years to when human manipulation became a permanent feature on the earth. What is the most profound effect of this change? Rapid wholesale decline in biodiversity. Mass extinction. Welcome to the Anthropocene.

Michael Soulé, the father of conservation biology and a Ph.D. population biologist out of Stanford, is one of a growing number of biologists adopting this new name, believing it most honestly addresses what is driving change on this planet right now. "We overhunt, overkill, we introduce exotic species, we start fires, we pollute," Soulé said. "Our sins go back hundreds of millions of years—selfish adaptive impulses and behaviors. You could say any living thing has them, but it matters that we are at the top. This is the first time in the history of the earth that one species has dominated and altered all the ecosystems in the world causing global crisis and extinction."

This new epoch, Anthropocene, has never sat well with me. It seems too early in the game. You can't name a geologic period while you're still in it. You don't know how it might play out, if it is even true in the long run. This could just be a spike in the earth's record, a sudden isolated anomaly. I am not ready for us to take the earth by name. I fear it will be embraced as a Hail Mary pass and we'll just keep running, never looking back, a future of nothing but humans and our artifacts, everything else be damned.

"I kind of agree with you," Soulé said. "But I think you're in denial about it. I don't like it either. I don't like the idea that humans so dominate the world that almost everything else is irrelevant. I hate it, I hate it. It makes me miserable and it makes me cry, but I'm a realist. I appreciate your optimism and holding on to this idea that things can turn around. I think they can, too, but I think the chances are really small and getting smaller every day."

"How long will this age last, though?" I asked him. "Is it worthy of a change in a geologic period?"

Soulé answered, "The markers that determine the changes are geological—changes in sediments, changes in minerals and compounds in the top layers of the earth, and that certainly has been happening quickly. We've destroyed most of the topsoils in the world and we've caused massive erosion. But it hasn't gone on very long. I think one of your points is that it's too soon to say this. Even though we are the most significant force in altering the geochemistry of the planet, it's just been an instant in geologic time. I kind of agree, but the Anthropocene does draw attention to the fact of how dramatically humans are altering everything, including the climate."

With a dismayed laugh, he said, "We hit seven billion people on the earth and what did all the media say? They celebrated. It just blew me away. When we reach nine billion, they'll celebrate again. It's that old instinct of wanting to dominate the world."

He pointed out that we are one of the youngest mammal species in existence, and in a very short time we've transformed the whole world. "The debate that humans and what we do are just part of nature goes on and on and it is really silly. It all depends on how you define nature. Hitler was part of civilization. Was Hitler good because of this? Are humans good because they came out of nature? You have to look at the consequences of what people do.

"The earth is dying," he said. "We could say humanity is nature's instrument to commit suicide."

We were sitting in his house in Colorado, dried flowers in a vase between us. Soulé ran his hand over his noble, bald cranium. "I'm getting old. I'm turning into more of a pessimist all the time. You don't really want to ask me these questions. I'm not very hopeful at this point. At least I'll die before you do."

He let go a bittersweet smile.

He was seventy-five. He had been struck by a heart attack a couple years earlier, which took some of the natural vigor from his face. Even though Soulé, founder of the Society for Conservation Biology, has done more to protect and even increase biodiversity than most living souls, how dare he say he was just going to pass on and let me and my children watch the world decline? I didn't want to take that for an answer. But Soulé was my elder, and a damn good scientist, so I kept my mouth shut.

No matter how Soulé looked at the evidence from climate to biology, it all pointed toward increased disruption of natural systems, and no such change comes without a price. "My optimism, a macabre form of optimism, is that the shit hits the fan a lot sooner than we think and forces us to wake up," he said.

"Will that stop the Anthropocene?" I asked.

"You can't stop it. It's here. The Holocene is gone, the megafauna are gone. A majority of the large, wild creatures will survive only in zoos by twenty fifty. Lions are disappearing from Africa, wolves and grizzlies from North America."

"I don't think the Holocene is over," I said, stubbornly.

"You're entitled to your opinion," Soulé said with a smile, as if he remembered once thinking the same.

Grasshoppers ate my clothes, the little bastards. I lifted a silk shirt, and it was riddled with holes big enough to stick three fingers through.

Angus was guzzling a bottle of water near his tent in muggy morning light. He glanced at my complaint. "You oughta keep that shirt. Hang it on your wall. 'Eaten Alive by Grasshoppers.' "

Angus snatched a grasshopper at the base of its thorax. He turned the armored creature upside down so he could study its scrabbling legs. Looking into the Samurai mask of its face, he said, "I don't think grasshoppers need to be admitted to the congress of creatures that need preservation."

"That's a little harsh, don't you think?" I asked.

"They're just not pleasant creatures," Angus said. "I don't see a need for them. I don't think the world would be especially worse off without grasshoppers."*

Yellowish with black chevrons on its jumping legs, this was the common yellow grasshopper, or differential grasshopper, *Melanoplus differentialis*, which eats everything it can and is known to swarm on occasion. Angus returned it to a corn leaf, where the grasshopper righted itself with the same armored expression, as if nothing had changed.

"In the post-alien-takeover world, I will have a position of oversight," Angus said. "I'll deal with grasshoppers at that point."

We left camp with enough water and food for the day. Looking for variation, for anything different, we mapped this six-hundred-acre plot step-by-step, following one row and then doubling back on another a quarter mile away. It didn't matter if what we found was human made (gaps where the seeder had momentarily failed) or if it was natural (a course of sand in the soil from some forgotten stream). Anything was worth noting, and we named landmarks we found: Mushroom Gap, North Swale, Sandy Swale, South Swale, Middle Tiling, Upper Tiling. We had our own little country laid out, figuring out how to navigate by combine turnarounds, alterations in row patterns, and patches of chaos where end rows were planted over long rows into snarled masses of cornstalks.

Angus called my name. I stopped. I couldn't see him.

"Where are you?" he asked.

---

* Grasshoppers account for eleven thousand recognized species worldwide. There are around twenty-four hundred genera within the grasshopper suborder of Caelifera (from Latin *caelare*, which means "to engrave in relief," referring to the sculptured integument of a grasshopper's exoskeleton). Grasshopper droppings are a significant topsoil fertilizer, while the insects themselves are a key food source worldwide for birds, spiders, mantises, and in some places humans.

"Over here," I said.

I slid crosswise through the stalks until I found him two rows over. We had dropped packs, carrying only water, easier to move. Angus had a new bandanna for the day—a bright orange giveaway, "Take Steps for Crohn's & Colitis." I had the same one as yesterday, and it smelled bad.

Under his bandanna, Angus looked like a librarian outlaw. We had been tunneling through a kaleidoscope for the last half hour, and it felt as if blood vessels in my head were going to burst. As we moved with the posture of bankers, our hands clasped behind our backs, shoulders forward, the sea of green had parted with our hat brims and sunglasses, closing in just behind us.

A crop duster buzzed on its way to other plots. We had asked beforehand if he would skip this plot while we were out. I wondered if he could see us moving down here, a tussling of corn heads like two odd whirlwinds out in the middle of a field. This field was in turn surrounded by several others just like it, exponentially enclosed by thousands of plots, an estimated three trillion cornstalks in this one state alone.

"It's a kind of wilderness," I said.

"It's a simulacrum of wilderness," Angus added. "It's an imitation. If you want to disappear, if that's your sole definition of wilderness, then we're there."

"Wilderness," like "biodiversity," is a word open to many interpretations. Sometimes for me it is just a place to fall off the map, a basic human need, at least for some. But Angus was right, all the other needs were not met. It felt like a dead wilderness, a domain of hallways and nothing more.

When I told Edward O. Wilson about this journey into the corn, he said, "It belies the feeling many people have that there is any kind of equivalency between what man has wrought and what nature has wrought and is being taken away. There's no equivalency at all."

Destruction and degradation of habitat are the leading causes of extinction, followed by competition from introduced species, making a modern cornfield a shining example of a worst-case scenario. Habitat loss worldwide is pushing to endangerment about 1,000 identified bird species, 1,600 amphibians, and 650 mammals. It felt as if we were in the very heart of this destruction. A light, blood-boiling breeze nodded through the uppermost leaves. Gashes of sunlight fell around us. Angus checked his watch, counting off in his head how far we were from our packs. The insides of my sunglasses were misted with sweat. I pulled them off and rubbed grease around with my thumb.

"I thought it'd be easy to get lost in here," Angus mused. "On the contrary, it would actually be very *hard* to get lost. It's simple mathematics, counting rows, knowing directions. I could write down how to get back to our packs and send someone directly there."

Sweat creased from around my eyes and ran down my face as I tried to remember how many rows we had crossed over and back. I was glad to have a land surveyor along, a person with a sense of scale. We were exploring in one direction and the next, still looking for anything alive.

Angus looked at me over his bandanna. "I don't know how much more there is to learn about this place."

"Let's keep going," I said.

"Whatever you say, man. This is your shitty idea."

I was adding up miniature flies and tiny spiders climbing their trapezes. Golden, long-legged orb weavers traveled through the upper corn leaves like gibbons in a forest. By now I had twenty-three distinct species marked down, a far cry from the thousands that would have been here in tall grass, but better than nothing, I supposed.

The sky looked like a cloud where we emerged out of a side of the field, our faces materializing from a wall of corn. On the other side of a barbed-wire fence stood a shocking deciduous woodland with bells of morning glories twining up branches. Whining like an engine, the trees were full of cicadas warmed up and buzzing high.

Pressed against the fence, I pulled down the greasy rag from in front of my mouth. "What is it?"

"Salvation," Angus said.

A creek curved through high grass, and beyond were drake darknesses and shades within trees. I glanced back at the riotous corn behind me, and I swear I could have run out of that place screaming if it weren't so terribly hot.

Angus slid down the fence to an old sign, letters nearly worn off:

RESTRICTED HUNTING AREA
GRUNDY CO. CONS. BD.

"It's a conservation boundary," he said. "Wildlife habitat. Farmer said something about it, remember? He said there's a bridge over a creek. There's the bridge."

An old plank bridge stretched across a hundred yards from where we popped through.

I started climbing a metal fence post, saying, "Let's check it out."

"And leave the cornfield?"

"Hell yes," I said as I landed on the other side. "I could use a break."

"What about staying in the corn?"

"It's just a little cheat."

Hell, I was desperate.

Angus gladly hopped over behind me. We crossed through hip-high grass under a washed-out white sun. Sprouts and grass weed came up through the old wood of the bridge, and below ran a graywater creek that seemed to be free from algae or water bugs. The Owens had warned us not to swim in the creek. They said it just wasn't healthy. On the other side, a grandmother maple twisted over the creek, its star-pointed leaves almost touching water, or what passed for water. Oak trees lined up behind it in rows, planted what looked like thirty years earlier as regular as corn. A mower had been through a couple months back, grass cut low where Angus lay down and cradled his head in his laced fingers as he sucked a grass stem between his teeth. As he waved away mosquitoes, he studied clouds passing through a humid sky.

"I always thought there should be a large whalelike animal that moves slowly about the sky," Angus said. "Creation is deficient that way. There should be more cloud whales and fewer fucking mosquitoes."

I moved on through unmowed grass toward a pond on the other side. Sedge came up as tall as my shoulders, and wing-tipped dragonflies cruised back and forth over my head. Grasshoppers snapped to the air, but they had different patterns from those in the ag drainage (big, green, two-striped *Melanoplus bivittatus* and long-winged *Stethophyma gracile,* the graceful sedge grasshopper). Black-winged damselflies with slender, iridescent abdomens rose in flurries as I passed to the edge of a pond ringed with bubbled string algae, the place feeling like a crowded zoo. Damselflies lit back down on leaves of bottlebrush sedge and the folds of my canvas pants. A lone, invasive bullfrog groaned from its bank grass shadows, fat Buddha half in the water.

Around the other side of the pond, I ducked into a thicket and vanished. Angus caught up not far behind. Stepping through umbels of small white flowers and sagging honewort leaves, I pushed brambleberried vines out of my way and emerged into wooded shadows.

"Now, this is different," Angus said as he moved in beneath a half-fallen shagbark hickory tree. "It's like a zoo for trees."

There was a sign, something that had been put here long ago, an informative plaque of some sort. I stepped up to it.

"Arborvitae," I said.

I looked up at a scale-leaf evergreen tree, cypress family.

"Tree of life," Angus translated. "Common ornamental."

"The name seems significant," I said.

"If you believe such things," he said. "Which I do."

Nearby was another sign, aquamarine lichens grown over the letters. "Red Osier Dogwood." That name didn't seem so auspicious, but it was pleasing to cup its globose cluster of pearl-sized fruits in my palm, not yet ripe and white as they'd be in a month. Beyond that was another. "Amur Honeysuckle." A name pleasing just to say out loud, sounded like buttery air sugar blossoming on a river in an exotic land.

"It's some kind of arboretum," Angus said.

"It doesn't get much visitation," I added. "I can't imagine not coming here all the time if I lived here. Do people not know about this?"

He said, "They drive to Des Moines for diversion."

Our footsteps mushed through leaves fallen a few seasons earlier, rotting into the ground. We split up, walking from sign to sign, some places grass had been mowed, other places left blanketed with black snakeroot and some kind of non-native turf grass. This place was kept so wildlife had a refuge, a product for human hunters. But it was at least a place. With long leaves slashing and bending and stalks spindling up from the ground, it is life that sets our planet apart, this copious, hungry reach for the light mixed with the weakened bend back toward the ground. The tree marked "Hawthorn" looked like a troll house, and "Juneberry" wove through itself with slender branches. "Red Mulberry" was only a stump. The sign for "Raspberry/Wild Strawberry" was broken, and I saw neither raspberries nor wild strawberries.

This woodland had been manufactured. There had been access to funds and a crew and someone telling them what to do to create wildlife browse, a deliberately designed refuge down to the ratio of species. Where an old snag tree looked soon to fall, younger trees around it had been girdled with chain saws to ensure new snags would stand in that same place. When the original surveyors came through in the early nineteenth century, they were mostly timbermen looking for places of lumber wealth, and they found Grundy County to be worthless, almost all of it tall grass or marsh. A woodland like this would not have been seen here. These trees had all been brought in. Even the creek was an invention, a reworking of drainages with buried tile and ditches. Everything here was by our hand. I was grateful anyway. What difference does it make? Was this somehow less natural? For this particular day, this was as good as it gets, birdsong through the canopy, sister butterflies with secret languages written in their delicate wings spelling out messages in the air. Each cell burned with its own small flame no matter the organism. There were so many.

More signs faded into the woods: "Box Elder," "American Elm." A trunk labeled "Green Ash" stood near the rest of itself, which had fallen and was moldering in the underbrush. The "Native Fern" sign was swamped in moss. The "Basswood" was strangled with wild grape.

Not every tree had a sign. Maybe one in twenty.

WHITE PINE.

AMERICAN HAZEL.

PRICKLY ASH.

Under the "Kentucky Coffee Tree" stood a single angle-white mushroom big as a cereal bowl, gills out for all to see.

Beneath the sheltering arms of a bur oak, Angus and I came upon an orgy of pencil-thick, six-inch-long millipedes crawling all over one another. They were all over the place, at least a hundred of them, as if they had rained out of this particular tree. They were umber colored down the middle and edged with a light sandalwood hue.

"What is this?" I asked, pacing out the moving mass on the ground, finding it went about twenty feet in either direction, but no farther.

"Clustering and exchanging genetic material, transporting it to other places," Angus said.

"That seems damn likely," I said, taking a twig and rolling one over to see its many small legs performing a wave as it curled into itself. Family, *Xystodesmidae*, widely known as black-and-yellows, though these colors were more muted. I rolled it back onto its feet, if you could call them that. With its segmented shields, it looked like something miniaturized straight out of the Carboniferous.

Millipedes are detritivores, meaning they live off ground litter, vultures of the plant kingdom. They are known to migrate in large numbers, moving mostly at night. The underside of this bur oak must have been a rest area on their way to someplace, perhaps just the right balance of edible detritus for a day's layover before pushing on in the heat of night. When dark came, I wondered if these

216

millipedes all lifted their heads, if you could call them that, no longer lounging upon each other, but all brought to attention, oriented in the same direction. As the millipedes drifted through twigs and grass, I imagined the scuttling sound of their progress on a secret journey like porpoises in a sea of detritus.

This felt like a refuge, but was it? I mentioned our serendipitous discovery of the place to Edward O. Wilson, and when I called it a forest, he said, "Was it really a forest?"

Wilson told me about visiting a similar place in his homeland in Alabama where a heavily disturbed longleaf pine ecosystem had grown back in. He said, "I thought it looked great. You know, it looked like a real, wild place." But a botanist with him, one who worked in the area, told him that no, it wasn't, saying that the floral diversity was actually quite low. Wilson said to the botanist, "Of course it will come back," referring to the longleaf pine forest. The botanist said no, he didn't think so, not anytime soon.

Wilson called this a post-climax ecosystem. He told me, "This is what happens when you have a catastrophic removal of a climax community and then you just let it come back helter-skelter. You will get something different, it might even look kind of natural, and it will persist for a very long period of time. It might or might not be replaced by the original climax."

What is the difference between an original, natural forest and something that comes back in after heavy disturbance? The ancient mosaic of different habitats that interacted with one another in the original climax forest is gone, replaced with a tangled mess that may require hundreds, even thousands of years to rise back to true diversity.

Wilson said, "I think most people, including ecologists, would not be able to tell they were in a post-climax or in one of these artificial assemblages that had been allowed to go on until it became a simulacrum of an older, more natural environment. To get into a patch of old-growth longleaf pine savanna is a revelation. It doesn't look that rich. You can't tell it until you start looking at the various plants and see the diversity among them."

What could I see without a botanist's eye?

Wilson told me, "Gradually, you begin to see what a true wilderness looks like."

We returned to the corn. We had to. Our packs were in there. Past the first step in, we were gone. Again, leaves closed around us as if the hands of hundreds of children were waving us through. With our heads down, hands clasped behind our backs, we marched between green sweatshop rows. Halfway to camp, we had to drop again, too hot to move, my head turning into a swamp. Somehow, humidity seemed to be higher and mosquitoes were following us, maybe all the way from the refuge. There was no breeze, no movement. Angus opened his mouth. "I'm fucking tired of being hot and sticky and dirty."

I had no pithy thing to say back to him.

At sunset we reached our packs. The molten softball of the sun fell through corn tassels. The air did not move. We sat in the tiling against the corn edge, neither of us wanting dinner, so we didn't fix any.

"No breeze," Angus said, his face dripping into his lap as he waved mosquitoes away. "I think this is a new kind of suffering from yesterday."

By early evening we were in our tents, no use staying out with the mosquitoes. I sat naked, not wanting my skin to touch anything, heat rash growing up the backs of my arms. Or was it a reaction to atrazine? Or maybe the first signs of gene transfer from genetically modified crops? If I moved too close to the tent walls, fabric reached out and adhered to me like a warm, sticky tongue, and I let out a pitiful groan as I peeled it off. The only sound was the whine of mosquitoes outside. I would have given anything for the air to move, for a breeze to saw through corn leaves and break the lead weight of this stillness. It wouldn't.

"Okay, two nights," I said. Angus listened silently, laughing to himself. I said, "Let's leave tomorrow."

Late that night, half awake on top of my sleeping bag, I heard the wind and woke. Air moved through the thin nylon walls of my tent. Corn rasped outside. My skin prickled. Was that coolness I felt? Then came a flash of light, a crack of thunder. Tap, tap-tap.

Rain.

In the morning, we packed up camp and moved through the baritone boom of thunder under a metal-gray sky. Blazes of lightning shot into the rows. Heavy rain set in as wind spun through the corn; roar of wind, roar of rain, roar of thunder, tassels dancing circles over our heads. Travel felt like swimming, reaching out with my hands to clear the way through slick and trickling leaves. This was yet another kind of suffering, not even cooler in the rain, our boots filling with water as we marched at a slap-thrashing pace.

We broke out of the corn and hit asphalt. A black man in bib overalls and an open raincoat was cinching down a load on a flatbed. He regarded us for a moment, two swamp creatures having emerged from the corn. He told us we were heading the right way to get back to Grundy Center. We started along the edge of the road feeling slightly refreshed, atrazine and the stink of anhydrous ammonia dripping off us and onto the ground in a warm summer rain.

7

Mountains Move

We're in the rapids now, heading for the falls,
too late to swim for shore.

—Bill McKibben

NORTHEAST TIBET

THE MONKS SAID next time you throw a rock at a dog, remember it could be your grandmother.

They said this because we were throwing rocks at dogs.

Every night at the monastery in the high fin-shaped mountains of northeast Tibet, the dogs had a wild rumpus and broke into our supplies. By day, the mastiffs lounged serenely, eyes half closed in utter contentment as they sheltered out of the rain under eaves around the bannered, whitewashed monastery. They only looked up to slap their tails happily. But after hours they became loud, wet mongrels snarling and stealing in the dark.

Yin and yang.

With no way to secure our bags of flour and sacks of butter and cheese we had hauled into this remote Khampa region—the territory of the so-called fighting Buddhists who gave the Chinese their greatest resistance—we could only guard them by sleeping next to them. Last night, the dogs jowled up about half of our yak-butter cheese, which was intended to be a large part of the fat intake for our coming expedition. Tonight, in the rain, it sounded as if they were ripping into our tsampa flour, nosing open metal lids, pulling out bags to get at the yak butter rolled up in it. And you heard the second they hit the frozen chicken. It sounded as if a dog bomb went off, a pack of deviled beasts snapping and yelping.

Somebody was up shouting, stumbling in boots. Rain came down

220

hard outside the tent, and I watched from my bag as a headlamp caught flashes of fur and thick-ruffed necks. The dogs ducked away from the light and skittered through the rain. The man had a few rocks in hand, winging them into the cone of light as he shouted, "Get outta here."

If Grandma was stealing our rations, she was going to get hit with a rock.

I lay back down in my bag. With sixteen of us packed into a canvas tent, the air smelled like a teenager's bedroom, fetid socks, humid armpits, and an open door letting in the steely scent of rain from outside.

The monsoon was coming hard. After a long, warm summer, moisture was being pulled in from the Indian Ocean and the Bay of Bengal nine hundred miles away. The height and mass of the Tibetan Plateau inhales weather. It drives the largest monsoons in the world. Formed by a collision of continental plates, it is evidence of a twitching, restive planet. This year was one of the longer monsoons on record. Every night it rained and wouldn't stop. In the pilgrims' tent, we wormed around in our bags to avoid drips and pools of water collecting on our thin, drenched tarps. Our purpose here was to navigate a stretch of river no one had yet attempted. Whether we were going to do it or not, we didn't know. Running off the steep northeast edge of the Tibetan Plateau, the Salween is one of the larger rivers to flow out of here and into Southeast Asia, and it was flooding. The last of its un-run stretches were being knocked off by rafters and kayakers every season, and within a few years these wild headwaters will be another known river. The summer had been warmer than usual. The plateau held the heat, and the heat pulled moisture from all around so that the monsoon lingered later than it should have. Whether we could do this depended on the rain.

We had arrived at the upper Salween in a caravan of beleaguered Land Cruisers and one growling supply truck. Five days ago we pulled up at the edge of a village where we found a river dangerously far into flood stage. The plan was to hit it then with rafts and kayaks, starting a two-week journey into cataracts incised into valleys twice as deep as the Grand Canyon. We were shocked by what we saw as trees floated down in the froth, big root balls rocking back and forth like monstrous heads. It was not the kind of river you would ever want to run. After three days of driving to get here, coming over a mud-slushed sixteen-thousand-foot pass, we weren't sure what to do next. It was mid-September, and every river coming out of Tibet was flooding like this, gritty water overtopping banks and swirling around piles of carved prayer stones.

At that moment, monks from a nearby monastery showed up on motorcycles and broke our dispirited gaze across the river. Word of our plans had preceded us. Crimson and cinnamon robes billowed around them as they pulled up to welcome us, a motley crew of fat, skinny, zit-pocked, handsome monks. In light, dappling rain (night was the time of more torrential downpours), they urged us to come back to the monastery so we could relax and think about what we were doing. They smiled and touched our arms, guiding us to get on the backs of their motorcycles.

They were concerned for us because just before we arrived, a little girl from the village fell into the river and was swept away. They sent warnings to villages far downstream so people could be ready with gaffs and long sticks to fish out her poor body. When they saw us come along with a ridiculous plan to do an obviously dangerous thing in that same river, they took note. Motoring us to their monastery along a mud-pooled road, they stopped before a white edifice strung with brightly colored prayer flags. The monastery seemed tucked into the undulating folds of the earth. No wonder transience is a bass note of the religion. These mountains curve like waves, one of the biggest mountain-building episodes in earth's history, erosion brought to an incredible spectacle, and in the middle of it all were these shiny towers and rippling flags. We were offered the pilgrims' tent while we considered our options.

After five days, the river was still up, and the rain was still coming.

"Are you awake?" came a low voice in the dark.

At my feet, it was my stepfather.

"Yeah," I said softly. Rain sounded like waves hitting the tent with every gust. I was still sitting up from the noise of the dogs.

"It's still raining pretty hard," he said. He sounded worried.

I somehow got my stepfather on this trip. This was the greatest geologic show on earth, and he is a geologist. It would have been a sin to leave him at home. A beloved man in his mid-sixties, he is an old-school field geologist, carries a Brunton and a couple hand lenses at all times. He was now "trip geologist." I had petitioned to bring him along, saying we needed a scientist. This was a John Wesley Powell sort of trip; you'd want the right people on your crew. The monks called him Pola, Grandfather, or, as he liked to say, Sage. He was the oldest one on the expedition. Married to my mom for fifteen years, he is a sturdy fellow, gives a strong twenty miles a day on foot. I knew him well enough that even in the dark I could see his round face and trim, conservative hair under a wool cap. He looked good with a beard, something he started growing after he married my mom. On this trip I took to calling him Old Pola.

He scooted up to his forearm and asked, "You think we're going to do it?"

"They're still debating, fifty-fifty," I said.

"Do you think it's a good idea?" he asked.

I didn't want to frighten him. He really didn't have a choice at this point. Whatever we did, he had to do also. I said, "It doesn't make a hell of a lot of sense to run it at flood stage, no. I'd want to see it drop about six feet. Then I'd feel more confident."

"But here's what I don't understand," Old Pola said, sitting up to face me in the dripping dark. "How do we know the flood doesn't actually make it easier by smoothing over rapids? Nobody knows what's down there, right?"

He was keeping his voice low, though in the rain no one else could hear him.

"Right, so it's a crapshoot either way," I said.

"Okay," he said.

He just wanted to make sure.

He crawled back into his bag and said good night.

. . .

223

Old Pola and I had run the Grand Canyon together in wooden dories. He was stiff-upper-lip all the way and I figured he could handle a trip like this. He knew how to hang on tight. We'd taken fifty-mile walks together back home in the Southwest, carrying rock hammers, maps, and iodine tablets. He'd been a professor at the Colorado School of Mines and then a prospector for the oil and gas industry, and he had an uncanny eye for the reason behind landscapes. He was always looking below the surface and seeing salt domes and invisible faults. We would sit looking at maps for hours, studying the various shapes and shifts on the planet's face, postulating over impact craters and upwarps, why this and why that. The earth was our puzzle. No smooth and cheerless marble, to us it looked like continents crashing and seafloors splitting open like the skin of a seedpod. It is this kind of bedrock action that drives some of the slowest and largest changes on earth. It is our reason for being.

I just hoped that this time I wasn't getting him killed. Omens were stacking up. Besides the news of the girl who slipped in and disappeared, we had heard a week earlier that half of a mostly Chinese white-water team of six had drowned not far from here when they wrapped their raft around an unseen boulder. We then got wind of an entire Russian team of eight dying while running another river off the north side of the plateau. All we heard was there had been a series of bad flips, and boats and bodies were seen downstream.

Warmer summers had been leading to delayed monsoon departures. Seasons were turning at different times, enough of a shift to get a bunch of river runners killed. This was the time of year river levels were supposed to be dropping off, but they were only going up. One of the earth's big clocks was changing.

The monks knew nothing of what was in the gorges below here, simply laughing and shaking their heads whenever we asked. To the Tibetans, big rivers are divisions in the land. Where rivers pass through gorges, they are not ways of travel.

When the monks discovered we had an indelible marker along, they grabbed it and started writing prayers all over our gear. They didn't explain why. When pressed, they only grinned as if the reason should be obvious. They spiked our helmets and oar blades with painstaking calligraphy, and they crouched over dry bags, raincoats, and ammo boxes, drawing sacred dragons enwreathed by holy sayings. Across my life vest, arced beneath my collarbone, was an incantation to walk in the light. It felt as if they were trying to gather us close, hoping something of their sensibilities would rub off before we decided to launch.

The Salween is the third sister to the Yangtze and the Mekong, turning south to form the border of Thailand and Myanmar before emptying into the Andaman Sea two thousand miles from its source. It cuts through the edge of the Tibetan Plateau, exposing the impact of the Indian subcontinent driving into Asia. This collision jacks up the highest landmass on the planet, the source of many of the world's great rivers, headwaters of the Amu Darya, Yangtze, Mekong, Salween, Irrawaddy, Ganges, and many others. Rivers coming out of the Tibetan Plateau flow as far as the Gobi, Bangladesh, coastal China, and Afghanistan.

We were up in the headwaters where swaybacked mountains edged up to eighteen and twenty thousand feet, a region known as the Tibetan Alps. In this heavily incised country, the monastery stood around twelve thousand feet in elevation. It was a green landscape dotted by farmsteads and barley terraces etched between landslide chutes where conifer-green woods crowded along rivers swollen by rain.

On the sixth night, we gathered in the monastery kitchen to cook over an open fire. Hands and knives moved back and forth chopping vegetables and chicken in the eclipsed light. The room smelled of steam and smoke. Headlamps came on as the last light faded from a grimy window as rain hushed against the roof. After we ate, Tripleader called attention.

Tripleader was twenty-three years old, a young kayak hotshot from Colorado who'd been taking first descents in the area all summer long. He'd been traveling here for a few years, knew the deeper, wilder places better than most foreigners. He took on this trip because he had the equipment, and there was money in it, everyone chipping in. He was a tall and long-muscled man with soft and inquisitive eyes. He was an expatriate, a river activist. His fight was to stop China from damming its every last river. Ours was to make a historic first descent on a stretch he had judged somehow viable.

"Me and the boat captains figured out a compromise," Tripleader said. "We'll start in the morning fifteen miles back up this tributary. It's never been run, looks like it has some class fours, enough we can work out the bugs, see how we do on this kind of water. That gives the Salween a couple more days before we have to decide."

"So that means we're doing it," said a theatrical young woman who had made finalist on the TV show *Survivor.*

"That means we're going to *try,*" Tripleader corrected.

"Fuck yes, we're going to try," said a burly Montana kayaker.

High fives went out, fists bumped over steaming bowls of food and a bottle of nasty Chinese grain alcohol we had broken out. I glanced at Old Pola, a midsized man like me with his shirt untucked, a shabby jacket, sitting on the ground with a bowl in his hands. He knew what this meant. Unless we made total idiots of ourselves on this tributary, you wouldn't be able to stop this momentum. He shrugged and went back to eating.

The next day, we drove up the side river on a road made of rocks and stilts. Hard black rock had been blown from a canyon wall to make way. Land Cruisers crept along, followed by our familiar supply truck, its flat-faced hood decked with rainbow prayer flags and Buddhist swastikas. Where we stopped to clear flood debris from the road, Old Pola was facing the blasted road cut, unlike everyone else watching the trucks grind through one at a time, the supply truck grabbing ground so it wouldn't fall into the river below. He had his hands on his hips and was looking up at the exposed rock wall.

"Dynamite blast," he said. "That's helpful."

"What do you see?" I asked.

"It looks like basement rock from down deep," he said. "It's all whopper-jawed. Look at those folds. Now, that's a taffied metamorphic. It's been down there getting cooked for a long time." He fished a hand lens out of his pocket. He called it Paleozoic schist, pointing out where it was rippled with bands of pale, crystalline pegmatites. "Rich in quartz and hornblende," he said, eye drawn to the hand lens.

"Where'd it come from?" I asked.

"It's the bottom of Asia," he said. "Pushed up by India."

As if, of course.

When the rock won't bend fast enough, earthquakes split through the crust. The majority occur on the Tibetan Plateau, where they spring like corn popping in a pressure cooker. Many strike the Indus margins of Afghanistan or inland China, or hit Kazakhstan or Iran. You can trace death tolls and monetary damage across the planet along lines and splinters as if there were a webbed blueprint of earthquake centers laid over the globe. These boundaries—Tibetan Plateau, Pacific Rim, Mid-Atlantic, and Pacific—are the contact lines between plates. Not that they are the only places. Even along sleeping faults far from collisions they snap every now and then; Washington, D.C., was hit with a foundation cracker in the summer of 2011 and it is far from any colliding plates. But the big ones you find at impact points. In 2011, Japan took a 9.0, followed minutes later by a tsunami that lit up a coastal nuclear power plant like a Roman candle and devastated towns, leaving nearly sixteen thousand dead. This happened at a subduction zone where the Pacific Plate runs below Japan, causing shifts of up to a hundred meters on the seafloor all along a fault line a hundred kilometers long. In 2010, I left Santiago, Chile, within hours of an 8.8, and the year before an 8.1 struck on the seafloor near Samoa. The big earthquake of 2008 hit outside Chengdu in eastern Sichuan, China, which came seven months after I flew out of Chengdu to Lhasa for this expedition. Earthquakes are not world-enders in themselves, but they tell of a naturally unstable planet down to the core.

In the scientific parlance of their destructive capability, these larger earthquakes are known as mega-thrusts, which tend to happen once or twice a year and almost exclusively at plate collisions. One that shifted beneath the South Pacific in 2004 was intensively studied worldwide, and seismologists saw a single rupture open for a thousand miles at five thousand miles per hour. The tsunamis from that strike killed nine hundred people on a variety of small islands. Likely places for mega-thrusts, within a several-hundred-year window, are Tokyo, Seattle, Santiago, Port-au-Prince, Tenerife, and Kashmir. They all lie at active contact points.*

These events happen because the earth's skin is as thin as an eggshell, riding atop a hot and welling interior. The eggshell is cracked into seven major tectonic plates and a handful of lesser pieces and microplates. These are in constant motion, grinding and colliding and pulling apart. The action ensures no landscape you are familiar with will ever remain.

Working in the American West, Old Pola used to take gravity measurements recording very slight changes in how hard the earth pulls downward in different places. He was fascinated to see gravity rising and falling, marking various densities of rock underfoot. What really struck him is how gravity changed not just from place to place but from time to time. He was seeing different readings by the hour. Densities were not staying put. He envisioned plumes and plunges under his feet, everything moving.

If Tibet were a glass-bottomed boat, I would have been looking into glowing, grinding pieces of earth below. Teeth of semi-molten gears were passing through one another as the earth's crust groaned upward, a slow volcano more than a thousand miles across where instead of lava there was solid rock.

Old Pola wrapped a knuckle on the rock face. It had verged on

---

* Studying the elastic, rock-rippling strain beneath the Himalayas and southern Tibet, the University of Colorado seismologist Roger Bilham concluded that there are far larger stores of energy than previously believed. He said quakes beneath the Himalayas were in the last thousand years much larger than any recorded in the past three hundred years and will be experienced again. Leading a study on a potential high-magnitude quake in Kashmir Valley, Bilham and his colleagues wrote, "The quake would be 200 kilometres wide as against 80 kilometres predicted earlier. The zone would encompass the Kashmir Valley, including the Srinagar city with its 1.5 million strong population. If slippage occurs over a length of 300 kilometres, as is possible, a mega quake of magnitude-9 is likely to occur. Given building codes and population in the region, it could mean a death toll of 300,000 people."

becoming mantle, nearly melted back into oblivion before it turned around and rocketed back to the surface. He looked back, seeing that each vehicle made it through the debris and rushing stream.

"I guess that means we're really going to do this," he said.

He headed for the line of waiting Land Cruisers to cram back in with a driver and five monsoon-smelling river runners musking with excitement. Within an hour we had hauled the show to the river's edge, picking a grassy place thronged with shore willows as our put-in spot. Rain cleared, and even slices of sun broke through. We took it as a good sign. Villagers and yak herders were down at the water watching as we rigged rafts and threw in dry bags. Motorcycles came riding in with families to see, four and five to each bike. The people touched everything. I found them petting my arms, amused by my hairs. Some stood in disbelief over the green-red-blue-tubed kayaks set in a row. Pulling on brightly colored dry suits and buckling our helmets, we must have looked like space people. We planned to just run a couple miles down the river, past a tiny village, and look for a place to camp. Tomorrow we'd get serious, hit some white water, see how we worked together; some of us signed up previously know-ing only a few on the team—an expedition by association, that is, a crapshoot.

I was on the first raft to push out. Starting on the oars was the trip doctor, a tall, sternly handsome ER technician from Colorado who would be the designated driver in any of the big rapids. "Nice and easy," he said as he palpated the current, spinning our raft into it. "Nice and easy."

The tributary ran through an iron split at the bottom of this steep, wooded valley. As currents jerked one way and the next, I leaned against the rubber tube peering up. The world changes when you start on a river. In a matter of seconds you have a new pace. High rock out-crops moved past each other as I marveled at the ten-thousand-foot depth of the place, a clean view of the basement rock, no dynamite needed. As I watched the taffied black bedrock fly past, I could see clearly the force involved between continental impacts, where pres-sure buckling underground goes up to ten thousand tons per square inch. We were entering the tectonic fist of the earth. I waved back at Tibetans above us on high-cut banks. They wore hats and skirts, and some had long, dark, red-beaded hair. Lone figures came out on outcrops much higher, word of our voyage spreading.

They thought we were crazy.

About a mile downstream, we proved it. The other three rafts were out of view, still rigging onshore and pushing in one by one. A Yellowstone-born woman named Brandy shot out ahead of us in her kayak. The water was squirrelly, its surface covered in tight eddies of all sizes. Even the smoothest water had strong fight to it, and Doc was having trouble finding any one current to follow.

Brandy flitted from place to place in her swift orange kayak, testing her arms. Her father was a legendary Yellowstone ranger, and she had been carrying a bear pistol since she was eight, a short, wiry, ferret-like woman, unflappable, it seemed. But she looked a little flustered by seething eddies that kept boiling up. Trying to skate the edge of a large one, she took a hole, and we saw her double-smacked by a pair of waves. Her kayak dumped.

Great, I thought. Ten minutes in and we have our first dump, haven't even encountered a rapid yet. We watched for a few seconds and saw nothing but the bright plastic belly and her paddle blade thrashing up from below as she tried to right herself while the kayak spun on the eddy line. I recognized the thrash. This cold water was more powerful than I had thought. It kept shoving her down, not a gap for her mouth or nose to break the surface.

"Come on, Brandy," Doc mumbled. "Get up . . . get up."

She pulled her cord and swam to the surface, gasping as the surge sucked half of her back in.

"Swim for your fucking life, Brandy!" Doc yelled as he knuckled his oars to catch her.

He had some drama in him, a Marlboro character with a big voice, but as far as I could see, Brandy needed to breathe. As we ran down alongside her, Doc's right oar blade caught the river's back sweep, and we shot into her eddy with the force of a car accident. While I half fell into the raft, Doc stayed on the oars, cursing as he strained toward Brandy, currents pushing us together and away. She swam doggedly, and once she'd made a reasonable ten feet from the eddy fence, she went limp, her mouth sucking hard at the air. We swept in behind her. I grabbed fistfuls of life jacket and pulled her in. She was small, which was good because she didn't have much kick left in her. Too early in the trip, I kept thinking. What kind of crew was this? As she keeled over coughing up water, her hands shook against the raft frame.

"Underestimated that one," she said, half in apology.

She wasn't used to being dumped so easily. The water had more muscle than she anticipated. Typical Montana boater, she was accustomed to pinball rivers, fast and loud with lots of boulders to hit. This was very different water. It had a penetrating brawn, not Rocky Mountain material, but Tibetan, the biggest steep water on earth.

We pulled her capsized kayak up and planted it across our frame, oars abandoned for the moment as we drifted close to the left shore. People had already come running. They poured down from a mud-wall, wood-roof village notched into the mountainside. Many of them wielded sticks and branches. One man carried a long wooden staff with a crook in the end. This is how they catch the bodies of those who drown.

As we drifted near the man with the staff, he reached out, knocked its crook onto our aluminum frame, and pulled us in. While we waited for Brandy to regain her composure, I glanced over my shoulder at about sixty Tibetans, and every set of eyes I met was filled with concern and warning. Men made death faces, tongues hanging out, and pointed at each of us, then pointed downstream. An old woman bowed toward me, her hands fused together in a desperate blessing. I did not know what expression to return. I gave her a polite, uncertain bow. I wanted to touch her shoulder and explain that it was all right. We were river people. But she would have grabbed my arm. Those around her would have reached for me. I would have been pulled back to shore.

· · ·

That evening we were a goat rodeo, ropes thrown and tangled, orders shouted as we barely caught a shoreline for camp. River right didn't have much to offer. Nor did river left. We were in a break in the canyon barely grabbing enough bank to hold a grove of evergreens. Some rafts missed the narrow landing or bumped each other out. A bumbling mess of oars swinging over heads ensued. Kayaks tried to get out of the way, squeezed between us as if between beaching whales. In a ferocious downpour, we cinched the boats tight, put up tents, and boiled a yak-meat supper. Conifers drooped, their bowed arms hanging above us in dark, misted steeples. Tripleader, with almost nervous animal eyes, looked uncertain. He kept saying that we'd see tomorrow.

All night it rained. The tributary hissed with gray sediments, mountains neatly disassembled grain by grain and carried downstream.

The Tibetan Plateau is why there is a deep rainy season throughout Southeast Asia. But its climatic effect is much larger than monsoons. It is a planet changer, this massive, high-elevation presence responsible for the multimillion-year cooling phase the earth has been in since not long after dinosaurs perished.

The planet's surface undergoes constant horizontal and vertical displacement. Landmasses rearrange and demolish old climatic regimes. Geochemistry of air and water changes globally.

Raymo and her early colleagues pioneered the uplift-weathering hypothesis,. They surmised the rise of mountains causes a decrease in worldwide temperatures. As mountains rise, fresh rock is exposed and bonds with CO_2 in the atmosphere. The rock erodes, carries its captured CO_2 down rivers to distant deltas, and buries it in carbonate rocks at the ocean floor where the gas is no longer available to warm the planet. When something as massive as the Tibetan Plateau comes up, it acts as a long, slow scrubber of CO_2, gradually turning down the world's thermostat. Short-term global warming aside, the planet has been in a cooling trend for at least forty million years. This is as long as the Tibetan Plateau has been rising, paving the way for ice ages and the current comfort of the Holocene.

"Tectonics controls everything about earth's climate," Raymo says.

"It controls the background state of the greenhouse gases; it controls the rate of mantle degassing; it controls the rate of CO_2 consumption. You can't separate long-term climate change from tectonics."

This happens over tens of millions of years, nothing we would see on a human timescale. Long-term is the key. Mountains move slowly, setting the longer course for the planet. Over the geologic history of this earth, climates have heaved one way and the next in response to the changing shape of the earth's surface. The fifty-million-year rise of the Himalayas and its mother mass of the Tibetan Plateau is what has ultimately allowed for ice ages and the singsong climate of the world as we know it. Raymo says, "Without the Tibetan Plateau, the background levels of CO_2 in the atmosphere would be much higher, never cold enough for ice sheets to survive through the summers."

As Carl Wunsch had said about seas acting on all timescales, the same can be said about tectonics. The collision of plates lets out volcanoes and throws earthquakes. It gives rise to monsoons, spinning global weather patterns, giving birth to rain shadows and deserts. Raymo says, "The elevation of the TP creates huge downstream meanders in the jet stream, which is why we have cold-air Arctic outbreaks over Canada, propagating over the entire Northern Hemisphere. Nothing can compare in scope in terms of present-day rapid uplift, the amount of erosion occurring, the Himalayas and Tibet are by far the largest contributor."

This planet has a restless heart, its interior swollen with heat turning over itself, crumpling and uncrumpling the surface over tens of millions of years, constantly unleveling the playing field.

Tectonic movement does more than just change atmospheric chemistry by sending up mountains and drawing down CO_2. It alters ocean currents by changing the shapes of continents. About four million years ago, the Isthmus of Panama lifted from the sea connecting North and South America, which divided the Pacific from the Atlantic. This was followed by the onset of intense and cyclic Arctic glaciations that have lasted up to today. Raymo and many of her colleagues have argued that the isthmus formed too early to be directly responsible for Northern Hemisphere glaciation. Others believe the closure of the Central American Seaway could have at least set the stage (along with orbital variations) for recurring ice ages by changing salt and heat transport between oceans, and intensifying the Gulf Stream, which carries warmth and moisture to the North Atlantic, creating conditions in which thick, permanent ice caps could form.

Early in the morning, the rain stopped. Heat left over from summer was letting go, allowing the monsoon to relax slightly, no longer pulling so hard on the Indian Ocean. We crawled from our sopping tents, surprised to see wheat-colored light nicking the highest mountains. It was the first morning sun we had seen. As if we had unknowingly positioned ourselves in the very eye of the canyon, the sun slanted down and struck us. We hung tents and sleeping bags in the warm light. We drank tea. I sat in the grass and read some Billy Collins I had brought along, an easy paperback suited to the frightening brevity of life. I felt for a moment as if we were lounging gods. This was the first drying we had been allowed in several days. Soon we suited up and were on the water, turned instantly mortal again in the roiling current. An old woman had come out to watch our departure from about a hundred feet up the other side. As we swept into the current, rafts tight on each other, kayaks flanking and on point, I looked up at her and noticed a flash of reflected sunlight swinging around her hand. She was spinning a prayer wheel.

As morning clouds parted, the day became beautiful, prismatic, everything bright. We were a flurry of kayaks, three catarafts, and one banana-yellow eighteen-foot self-bailer. The banana boat was captained by a surprisingly tall woman from Berkeley with lizard-like musculature leaning into the oars. In her late thirties, she had laser-pointer eyes. I don't think I ever saw her blink in the week we'd been together. She had just finished her dissertation at the University of California, Berkeley, on the politics of Chinese hydropower, and this was her first trip out, a celebration of sorts. She had also been moonlighting as a cello player for a punk rock band in the Bay Area, and I had found her drunk in a Tibetan karaoke bar in Chamdo singing as if there were no tomorrow. She had a good read on the water, shouting orders as the other two of us in the raft flew with paddles in our hands—*back paddle, draw, hard left, hard right.* We dropped through ledge-fall rapids as kayaks ahead signaled directions back at us for what we couldn't see, *left, right, down the middle.* When our flotilla slid into shore just above the confluence with the Salween, we looked like a flock of swans. Bow jumpers leaped to land and pulled ropes tightly. Kayaks nosed up one after the next.

I nudged Old Pola and said, "See, it matters when you go with pros."

If you want a calmer river, look for coolness beneath the earth's crust, not hot like the Andes or Great Himalayas, but the Midwest where winds the Mississippi, or the muddy Amazon of eastern Brazil. Temperature is what drives the mountainous twisting and turning of more complex terrain. Without adequate heat, this planet would look more like the moon or Mars, a one-plate planet. Or with too much internal heat, we might have ended up more like superheated Venus, which has more active volcanoes than any other planet in the solar system.

We walked to the edge of the Salween, a river cutting against the steep edge of the world's largest, highest plateau. The river's broad, roiling surface had barely changed since we last saw it. The river was still in full flood.

Tripleader crouched at its edge.

"It's come down a couple inches," he said.

That wasn't the best news. We wanted it down six or seven feet.

With his long arms draped over limber knees, he sat on his heels as he peered downriver like a crouching frog. "Better than nothing," he said.

The place he had chosen to crouch was atop a pile of old prayer stones, a rough shrine with its toe in the floodwaters, river thick with mud as it hurtled by. The prayer stones gave him a little elevation, a couple feet of extra vantage. There was no insult in him using the pile. I'd learned already that they were meant for use. Truck drivers bringing supplies down these far-flung roads pulled out these intricately carved stones and poured them in ruts. Monks had told me this is what they are for, to usher prayers into the world, to keep them moving.

While most of the crew talked about the sizes of rapids and what such a flood might do to them, Tripleader, who was purely Anglo, had spoken more of karma, destiny, those sorts of things.

"It's important to do the right thing," he said. "The river knows."

Tripleader had been in the country for a few years off and on. His father was a Grand Canyon river ranger who had brought him out here young, seeking wilder water. This was his first time as the head of an expedition without his dad, his first real assignment. The crew from Montana had sought him out and put up the money, and the expedition was on. Now we waited for his decision.

As his hands absentmindedly tipped to the ground, fingers feeling the carved lettering of prayer stones, I wondered what he was imagining downstream of here. Towering gorges with the skeletons of our raft crushed against boulders, survivors huddled on a box-canyon floor, the others lost, bodies never to be seen again? Or did he see a flawless ride, high fives, and the shouts of champions?

He stood, wiped off his hands, and said, "We're going to do it."

The Salween proved oceanic. Floodwater glided into huge, mud-slick waves. No more Tibetans came to watch from shore as mountains shouldered into the river. The rapids we hit were big but passable, sequences of swirling pools between sudden drops, a rhythm of big water where all we had to do was set up, get some momentum, and smash through. Doc shouted to Old Pola in the middle of a rapid as water crashed across his face, "Stay in this boat, stay in this boat." He wanted this to be a historic first descent, and you could hear it in his voice.

We glued our eyes to the opening bends of the canyon, every frame of river registered as it came into view. The only maps with us now were Soviet made, from the 1970s, and so far were not proving wholly correct. Certain mountains were missing. Others were where they shouldn't be.

Gyalmo Ngulchu, locals call this upper stretch of the Salween. It means "Tears of the Beautiful Princess," named for Gyalmo, a celebrated princess from China who married a powerful Tibetan king several hundred years ago. It was a historic marriage, and to fulfill her role, Gyalmo traveled from her home to marry the king deep inside Tibet at Lhasa. The journey brought her through the teeth of the earth in eastern Tibet, where her trail would have been narrow as she was carried in her divan. Her entourage followed with statues and scrolls, the dowry of her country, payment for a key political alliance. You can imagine her transport rocking back and forth along edges over which she dared not look. Her hand must have reached and pulled open the blind as this young, decorated woman looked out. Accustomed to grain fields and open country from her home in western China, she must have found this gut-wrenching. Overwhelmed by the sheer monstrosity of the landscape, she cried as she passed through here, and, according to local legend, her tears became the river, the Gyalmo Ngulchu.

We sank into a dark, near-vertical canyon, where by the end of the day storm clouds had rolled back in. Slopes were grazed by rock outcrops and shields of forest. The river was faster than we had anticipated, making it hard to stop. Trying to get off it would be like hopping off a freight train. But we had to. Light would be failing soon.

Coming around the bend, hoping for some high-water gravel bar still exposed, or at least a little pinch of forest, we saw exactly what we did not want. The right-hand mountain face was gone, caved straight into the river. A bright gray landslide had taken out a thousand vertical feet of mountain, our shoreline gone and replaced with an incline of boulders and chalky debris dropping straight into the river.

Oars and paddles slapped the water hard. We caught a skim of the last possible boulders stabbed out of the riverside. Jumpers hit shore among busted timbers, where they wrapped lines as fast as they could. Ropes went tight, tying into what had once been forest and was now splintered trees chopped off at the knees. Anchored in, stationary for a moment, boat captains leaned against their oars and gazed downstream. Fresh, flood-buried boulders had yet to be hydraulically seated, and the river chundered shore to shore over the collapsed mass, circus-tent waves crashing into one another. You couldn't even call it white water, it was so dark with mud. It wound around the bend and vanished from sight. Pegs of hail began popping off our helmets. We were all thinking the same thing: water always looks worse in a storm.

The landslide left an impressive scar, most of a mountain's face swept down. This was what Raymo was speaking of, fresh rock exposed by uplifting mountains. A layer of carbon-saturated surface rock was off to the ocean, and a fresh new skin was exposed to start the inhaling process over again, slowly drawing CO_2 from the atmosphere and sending it away. Our concerns were more immediate, though. Rather than changing climate, this uplifting plateau was directly affecting our day.

The first decision was to ghost the rafts—push them in unmanned and collect the carnage downstream. It just looked too big to run. Normally, you'd try to line boats down a rapid like this, easing them along with ropes from shore, but there was no shore, just a landslide dropping straight to the river's violent cut. If you knew what was around the corner, and it didn't look like a churning death fall, sure, you'd run it. But if that wasn't the case, you didn't want to find yourself running blind and already wave bashed into the biggest rapid you've ever seen.

A man with the lean face of a wilderness junkie, an Outward Bound guide back in Montana, scrambled out to the sheer, loose edge of the landslide. Every rock sliding out from under him tumbled down into the hissing, muddy water. He leaned out as far as he could to see around the corner over the scathing roar. We watched him and waited for a signal. He didn't look convinced of anything. Walking back to the rest of us sidestepped into the wall of the landslide, Outward Bound said, "It just keeps going. I can't see the end of it."

Talk went on, ghosting starting to sound like a nightmare. Outward Bound shook his head and said, "If we're just going to ghost them, I might as well be in a raft."

That was the decision, then. Run the rapid one to a raft, no more. Everyone else would scale the edge of the landslide and walk it out, life vests and dry suits tight in case someone fell in. We'd meet downstream at the soonest possible landing.

Walkers grabbed food bars and whatever they could, zipping them into their dry suits in case this ended up being a much longer and more harrowing day than we were planning.

Old Pola approached. "How are we doing?"

I looked up from a brick of military protein ration I was tucking away. "Just keep your helmet on," I said. "And go nice and slow."

I knew he didn't like heights and perilous footing. Who does, really? Especially in the rain and above a thundering river. I've long admired the man for his resolve, his steadiness. I was more worried for him than for me as I chipped in steps and crept along in front of him. We dug with our fingers for balance, gripping cold, wet grit where a stream once poured through, now obliterated. We had rope, but no use for it. This was no ice crossing or high-pitch belay. It was newly cleaved boulders balanced in the barest repose. Nothing could be used as an anchor. Crossing sheets of pulverized rock and half-slurried sand, we reached a shattered outcrop of bedrock that blocked the way. Survivor Woman unloaded the racks of climbing gear she'd brought along for just this circumstance. Old Pola didn't trust her, and he kept looking at me and I kept shrugging. I didn't trust her either. She was a bit too enthusiastic for my taste, slinging out rope as she began climbing and setting anchors as if we were on some life-or-death mission.

I glanced at the photographer from Arizona, who had been sent by a magazine to cover the expedition. He was glancing at me. We both had children back at home, he with a daughter, I with two sons. His daughter and my youngest were both just learning to walk. We had a shared reason for making it out intact.

Rope in this situation was not our style. The photographer and I were Grand Canyon types accustomed to navigating steep river-bottomed rock slides. Our inclination was to go up and across at whatever opportunities revealed themselves much higher, above the unstable first hundred feet of river cut. I patted Old Pola on the shoulder and quietly said, "We're going to fall back and take a bypass up that way. You'll be safer down here."

Old Pola turned so he was speaking to the ground. "With her?"

I pointed up. "Your choice."

Above rose the sum of Old Pola's fears, unprotected heights and crumbling rock, while below was the furious clamor of the river.

"Just do exactly what she tells you," I told him. "And don't get hit in the head with a rock. You'll be fine."

I was starting to feel like a bald-faced liar to the man, but what choice was there? It was too late to do anything but head downstream.

Ten minutes later, the photographer and I were high in the tiers of erosion, kicking in quick footholds, rising like long-armed monkeys. We stopped and looked down to see the progress of the rope team hundreds of feet below. Survivor Woman was cinching a harness around Old Pola's crotch, which made me shake my head with a bittersweet laugh. All I told him coming into this was that there was a chance of this expedition going hairball at any point, sign on the dotted line, your life is in your own hands. On the plane flying over, bumping through monsoon clouds and dropping over Tibetan river towns half drowned with floodwater, he had said, "If something happens to me, it is not your fault."

Nothing would happen to him. I just believed it. And I hoped believing was enough.

The photographer and I stayed in back of the rope climbers so we wouldn't send rocks careening in their direction. We crouched on our heels and watched the first kayaks slice through the rapids, paddles flying as the photographer fired off shots, switching out cameras from his duffel.

Below, Old Pola was arguing with Survivor Woman. He wanted to watch the boats run their rapid, and she barked at him to climb. None of this could we hear over the clamoring river, but I could see it playing out below as people gesticulated and each kayak disappeared around the corner followed by the train of rafts bucking, shoving, twisting. They continued upright and slipped out of view.

"That didn't look so bad," the photographer said, lowering his camera. "I give them a mile before they figure out to pull over."

He turned back to the slope and chopped his way up about five feet over from me so I'd be out of the way of his rocks. I stayed for a moment, weight on the balls of my feet. My dry suit already felt like a bucket of sweat, tight under a cinched life vest. Up here, it didn't matter if I fell. I'd be half dismembered before I even hit the river. I fisted open the rubber gasket snugged around my throat, letting in raindrops and cool air.

It must have been impressive when it slipped. Hundreds of thousands of tons of rock let loose at once, and forests fell over themselves as if a cloudy rug had been pulled out from under them. The river must have gone absolutely mad, dammed and breaking through all at once. Downstream would have been a mess.

Maybe it was an earthquake that triggered this landslide. It didn't have to be, though. It looked as if everything around us was ready to go, the river cutting down through earth like a snake whipping through tall grass. This is how mountains move. Not only do they rise, but at the same time they fall. Their fathomless weight is carried away one monsoon after the next, boulders turned to grains, grains to mud, and mud into fans of densely packed sediment off the coast of Thailand. Uplift and ensuing erosion turns the planet into a sort of perpetual motion device. When the climbers were done below, I sidestepped to the high, far edge of the landslide. Beyond crushed and splintered trees was a steep and mossy forest. Booting through the debris, I moved from yin to yang, out of a sheared-off wasteland and into rhododendron branches tangled among each other. I pushed in steps along a spongy floor of rain-swollen lichens, green closing behind me like a curtain, exit concealed. I wondered where everyone else was. We had violated one of those wilderness rules: don't lose each other in times of crisis. But it couldn't last long. Downstream was the only direction. We'd have to find each other again.

Mist wet my lungs, smell of enchanted wood. Boulders balanced like the heads of lost statues, remnants of landslides long ago. I climbed over them, seeking the more stable angles, the ones with the most lichens covering their faces. As I grabbed my way through tree trunks and jumped down mossy flumes, it sounded as if I'd gotten beyond the rapid. The canyon was quiet now, only a distant roar behind me. I turned down through purple-coned spruce and rhododendrons big as cottonwood trees, sliding and landing till I reached the brisk swirl of shore. Another quarter mile downstream I found the rafts tied to one another, and kayaks pulled up on rain-glossed boulders. Everyone had made it.

Tents went up in a downpour. A hunk of yak was reboiled in a pressure cooker, product of the People's Republic of China. Even in the rain there were smiles, looks of satisfaction. They said the landslide rapid was big and punchy but not over the top. It played into a nice outwash, easy to get your bearings again. The first gateway was past, and now we were in.

"Gravity, Craig," Old Pola said to me. "Now, there's a force to match tectonics. What goes up must come down."

Mountains rise, and as they do, they gather precipitation. Rivers run hard and tear the mountains apart.

Tectonic forces push up from below, while gravity—hence erosion—pushes down from above. Debris carried away by rivers and wind shifts weight from place to place, pushing down on deltas or landing in sand-dune seas that lean their immense heaviness into the earth, changing the pressures inside.* Even monsoons change the way tectonics behaves. When they have intensified in geologic history, they have moved more sediment, the weight shift causing

---

* The flow of global sediment through soil erosion is estimated to have been increased by around 2.3 billion metric tons per year compared with prehuman conditions. At the same time, by building dams and holding back river-carried sediment, we have reduced the amount of sediment that reaches the sea by about 1.4 billion metric tons per year. So, in a sense, we have become a tectonic agent.

a measurable reaction in the speed and direction of the Indian Plate as if the earth were a sort of geologic organism, a contained, self-changing system.*

Water is not the only vehicle. Wind blows so much sediment out of places like the Qaidam Basin of inner Asia that it shifts planetary weight, changing the tectonic balance between Asia and India. Paul Kapp, a geomorphologist at the University of Arizona, has been compiling ancient wind patterns around the Qaidam Basin, in the rain shadow of the Tibetan Plateau, where he discovered that when windstorms intensified about three million years ago, bedrock far below began to fold more markedly. Kapp calls the process "wind-enhanced tectonics."

This is the yin and yang of the earth, an energetic feedback. What happens below relates directly to what is happening on the surface and in the atmosphere and vice versa. Tectonics does not end at the ground beneath your feet. It is a dynamic system from the earth's interior all the way into the sky and back. The sediment we were riding down the river was adding to it. Over millions of years, river deltas grow heavier, and eventually their water-saturated floors dive into the earth's mantle. The water they carry (turned into a super-pressurized and almost plastic form as it nears the mantle, anywhere from five to fifty miles beneath the surface) literally lubricates the tectonic machine, keeping it running so that new mountains can slide up as if on melted butter.

---

* Giampiero Iaffaldano, a research fellow at Australian National University's Research School of Earth Sciences, produced a model along with colleagues in France and Germany looking at how monsoonal erosion on the Tibetan Plateau changes the behavior of plate collisions between India and Asia. In one model, intensified monsoons lead to increased erosion, at which point the Indian Plate starts speeding up and turning in a clockwise motion based on how weight is being redistributed on top of it. This matches what appears to have happened over the past ten million years as monsoons strengthened based on how high the Tibetan Plateau was lifting, and the tectonic plate beneath it began turning clockwise and picking up speed. Faster erosion in the eastern Himalayas appears to have reduced the resistance between converging plates, actually allowing India to move faster by about eight millimeters per year and in a slightly different direction. When his findings came out in 2011, Iaffaldano said, "Long-term climate change, or the natural changes in climate patterns over millions of years, can modify the motion of plates in a feedback mechanism."

The watercourse before us opened and closed as it inhaled into tight-lipped gorges and exhaled into soaring valleys. Wherever the river widened, shore boulders flew past with faces daggered in Tibetan lettering, "Om Mani Peme Hung," a prayer of compassion to all living things. These boulders were here to broadcast messages, speaking over and over to passing wind and water, adding to this procession of mountains falling down the river.

Morning clouds broke into silvery light coasted by plank-winged vultures looking for something dead. The rapids weren't so big on day two. They were just push and pull, eddy lines spinning us against our wills. We had to jump across our rafts high siding as the river boiled outrageously, sucking down one tube and the next. This was not a river to be easily finessed. It was far stronger than the tributary, big hydraulics. Kayaks not paying attention turned to swizzle sticks, sterns popped in the air as they struggled to paddle themselves back into a horizontal position. Maybe this was easier in flood, rapids swallowed beneath us.

As we rounded a bend, we came into view of a woman performing some small task at shore. She looked up, stood, and stared at us for a moment. With no further ado, she turned around and ran into a hut notched into the face of a green "Hansel and Gretel" glen. She did not even peek out the door again, and I imagined her fingers busily thumbing through prayer beads at the mere sight of us.

Three rapids, four, five. We skimmed curlers and broke through waves like dolphins. We scouted, argued, decided, ran. The photographer got a ride ahead and set up shots as we ran the bigger drops, but in the larger rapids there were no pictures. By the next day, several inches of new shoreline were starting to show. The monsoons were weakening. Sunlight glittered through breaking clouds and touched the horns of glacier-cut mountains. We pulled in at a forested creek, tied off, and turned over a kayak onshore to use as a table. We sliced salami and cheese and opened packages of cookies on the rock-scraped bottom of the kayak. Once the food was out and I snatched a mouthful of salami, I climbed up through the woods with my dry suit unzipped down to my waist, the rubber gasket pulled from around my neck so I could breathe more easily.

The river was moving too fast, and we could barely find places to stop. The trip would take half the time we had planned only because of the speed of the water. Whenever we landed, I took off trying to get a sense of where we were, what sort of geography we were entering. I scrambled up between bristle-limbed pines, pulling my way up by their trunks until I could no longer hear the river. The quiet was relieving. I am more of a walker, really. I do rivers on the side, but the way I prefer traveling is by foot. No paddles or helmets or terrifying U-turns into eddies while trying to pull over. I like running my fingers through needles and the soft sound of steps pressing into duff. I stopped, thinking I heard something. It was in the distance, a musical jangling. As I came closer, the sound continued, like chimes ringing in a prayer hall. I peered through the trees, expecting to see a humble house and a grinning bald man in a robe, a guru waiting for a follower to arrive. All I saw was sunlight through a clearing, and I moved to it.

Horses did not even look up. Fifteen of them, white and brown, browsed grass in a small pasture cleared out of the trees and filled with sunlight. They were hungry, freshly returned from a barley run, their withers still strapped with sweat from where they had carried loads. Each wore a copper bell, and each bell had a slightly different tone. As they nipped and tugged at the grass, the bells rang.

I stayed at the edge of shadows and crouched as the horses moved from place to place. Their manes and tails were neatly combed and ribboned with strips of colorful fabric. Big yellow teeth scraped up grass, leathery lips slapping as they chewed. One glanced at me for a moment, then returned to its task. The bells chimed and tinged, a small ensemble of musicians gathered for a recital.

José Rial would have liked this. Listening to the patterns of ice quakes in Greenland, he found a certain musicality in the randomness. The high, lustrous tones of the horse bells were arrhythmic, as if played by musicians with their eyes closed. I squatted at the edge of the trees and closed my eyes, too, listening through clusters of sound mixed with brief silences. I often think of the earth as having a cadence, a drumbeat tempo of ice ages and sea levels, civilizations rising and then falling. But thinking of Rial, I was learning that you do not set calendars and metronomes to the patterns of the earth. They are melodies, not drumbeats. It was music that we were moving through with paddles and oars, the music of centuries and millions of years. As the weather cooled toward autumn, the rains were beginning to soften, the draw on the Indian Ocean relaxing. The flood was subsiding.

This is the one planet in the solar system with plate tectonics; no small coincidence it is the only one with life, at least the copious shows of it we are used to.* Astrobiologists pondering the likelihood of life on planets beyond our solar system are hoping for arcs and knots of mountain ranges, signs of tectonic movement beneath the surface. The movement of plates sends surface material back into the planet and back out as if composting it, enriching it with minerals which are required for living organisms to thrive. Life changes atmospheric balances, breaks down rocks, creates soil, causes differential erosion by growing in some places and not in others. Tectonics and life appear to go hand in hand. Life requires change to exist, a surface overturned not too fast, not too slow, new geographies constantly being built on which life and evolution feeds.

This dance between life and the solid earth does not always go the way you might wish. Look back 250 million years to the end-Permian extinction, when the supercontinent Pangaea had crashed into place and was just starting the process of tearing itself apart again. For a moment, some fifty million years, the earth would have appeared lopsided, with continents jammed together all in one loca-

---

*Tectonics may be responsible for even human evolution. Accelerated mountain building from Ethiopia to South Africa over the past seven million years created a rain shadow, turning what was once jungle into an arid mosaic of woodlands and open savanna, a habitat that favored our ancestors.

tion, their impact creating an elevational uplift that would have dwarfed even the Tibetan Plateau (what is left of that is the Appalachians, metamorphic nubs eroding back toward almost flat). Meanwhile, there was only one ocean, which would have occupied most of the planet. Climate models based on this configuration show a sort of global doldrums taking over. With only one ocean and no gauntlet of landforms as we have today, multifarious ocean currents would have joined into one mass, a warm bath moving freely from the equator to the poles around the other side and back to the equator. Meanwhile, the singular landmass of Pangaea would have heated up, with huge monsoons dragged into the interior by its giant mountain range, initiating runaway erosion that has left evidence of massive sediment loads dumped out of Pangaean coastlines. And the uplift of this ancient mountain range would have cooled the planet, making it drier. At its most fundamental, geographic level, you can see a stage set for a mass extinction, the earth simplified, diversity of global yins and yangs reduced to one continent, one giant range, one sea. If there's one thing life doesn't handle well, it is a rapid plunge in any form of diversity, be it biotic or geologic or both at once.

We now appear to have a healthy balance of continents and seas, our mélange of ocean and air currents mixed with landmasses and mountain ranges of all sizes and configurations. But don't get too comfortable thinking in geologic terms. We may have hundreds of millions of years to play with in our minds, but it always comes back to this crystalline moment of now.

We were worried about the next gorge downstream. Stopping near the edge of a small village, we bought some beers and asked about what was downstream. Standing on a single-track horse and motorcycle trail along the river, a group of Tibetan men showed with their hands a waterfall. They tugged at our clothes and smiled for us not to do it. Our Tibetan translator was having trouble with their dialect, but the message was clear: They thought we were going to die. We gestured in return, showing them the size of rapids we had been through in the last few days. They shook their heads, insisting we didn't understand and that we would face a waterfall downstream.

But what did they know? This would be the narrowest, most inaccessible gorge along the Gyalmo Ngulchu. We figured no one had ever been in there.

We camped above the gorge, everyone a little nervous. So far, no raft had flipped. The biggest rapids took us to our limit and let us go. But Tripleader was pacing that night, beer in hand, and Old Pola took me aside, asking, "Do you think there's a waterfall in there?"

"I sure hope not."

As first light broke through clouds the next morning, I stood at the edge of the river looking down into folded shadows, hoping the river would go flat from here on out. I'd be happy if there weren't another single rapid. I guess I should admit rapids make me a little anxious. Just their sound injects a cold shot to the blood. I was getting tired of challenges, of my heart beating up in my throat the last few days, not knowing what was around any of these bends or how big the drops might be. Before we started this expedition, Tripleader had stayed up nights studying the next gorge on his computer using Google Earth, which showed a six-mile-long squiggle hamburgered by pixilation, its inside black with shadow, and no sign of any river at the bottom. He figured there was a drop of at least sixty feet in those six miles. Falling all at once, or stretched out evenly, we wouldn't know until we were in.

Breakfast was nearly silent. We went through the motions, tying equipment down, every strap pulled tight, life vests buckled and cinched. Even Doc was quiet. We pushed in, and the river took us.

Two men were gathering rocks on the other side and stopped to watch as we drifted by, kayaks fishing between the rafts. As the terrain steepened, we saw one last person, a woman sitting on exposed river cobbles. She did not wave. She simply stared as we swept into the dark mouth. I wondered about these women I'd been seeing along the river, hoping they were somehow good luck.

The first half mile the river swirled like stew in a cauldron. We seemed to be entering a mountain. At first, no rapids cropped up.

We started thinking we were going to be fine, just a busy, bustling stretch, but then the water fell. Shouts went out, crew jumping into position. Kayaks out front signaled, but they were giving different signals, and then they were gone. Water twisted backward, unable to fit all at once. Waves began overlapping, bucking against each other. I straddled a raft tube as my right leg twisted fitfully in the water where I leaned out with my paddle and dug. I ducked my head, and we were brick walled by a wave, raft bending almost in two. Water socked into my sinuses.

I was in the banana-colored self-bailer, Berkeley Woman on the oars. She blurted orders, spinning us around to take waves with her back, then her front. All I could see was froth, rock, froth, and the satin tongue of a massive wave that rose over my head and exploded onto us. It was instant chaos, rafts nearly running into each other, kayaks scattered like confetti. My paddle grabbed air, then water, then air. It was too fast, complete bedlam.

We slammed into the first eddy we could find, joining the flotsam of kayaks and rafts seeking shelter. The eddy was almost more dangerous than the rest of the river. Whirling currents threw us into each other, oars flying over heads as we ducked. The eddy sucked down raft tubes, garbling and spitting them back up as we leaped for the boulders, ropes tossed ahead and tied off.

The rapid was much larger and more continuous than we had thought, waves topping ten feet, a few approaching twenty. What was ahead? Tripleader and the burly Montana kayaker turned in to the current and shot downstream with a radio. The two kayaks popped in and out of sight, engulfed, spit back up, then were slung around the bend as we stood in a transfixing roar.

We'd wait for their report.

Not that we had any other option but going downstream. Even Survivor Woman couldn't get us out of here with all her ropes and chocks.

We were in a notch carved from the side of a chasm that looked to be a few thousand feet deep. I craned my neck to see steepled, monolithic heads cracked and leaning. The shore, if you could call it

249

that, was made of tilted boulders, and we waited on points and faces like mountain goats, thumbs hooked in our life vests. We listened for anything from the radio.

We didn't belong here. This was not a human place. It felt like falling into a bottomless space, one of those places where you realize the earth was not made only for us. Mountains had not erected themselves with our stride in mind. Why did we ever think we could do this? In the reeling and cracking of a very live planet, we were flesh and bones looking for a way through.

I could feel my heart beating through my life vest where a monk had written that prayer to walk in the light. I hoped it was enough to help get us through. No one spoke. I didn't even look at Old Pola. There was nothing I could say to him. In this kind of water, if you get thrown in, you're probably going to drown.

I didn't want to die here. I couldn't. I was no hero. I thought of my two boys. I thought I would claw my way straight through this planet to get back to them.

Three minutes passed with no word from the two kayakers. Then five minutes.

Berkeley Woman was dancing around on her toes, letting her head and arms flop around as if she were a puppet. She was shaking doubt and fear out of her body. Rafting was just one of her many hobbies, and she didn't have time on the water anywhere near the other boat captains. When she finished dancing, her face had taken on athletic resolve. She could have cut flesh with her eyes. Right then, I would have followed her anywhere.

Ten minutes went by and the radio remained silent. No one wanted to say it, but the kayakers either could be dead or were unable to figure out what to tell us. Either way, it didn't matter. We still had to go down there.

After fifteen minutes, the radio crackled with Tripleader's voice. He sounded out of breath. Or was it a bad connection? As we gathered around, I heard him say there was no letup in the rapid, but they had found a place to eddy-out on the left. He started into his instructions: "Left, hard left, and then way right around a hole you won't see."

Burly Montana interrupted in the background. They were debating about the run. Tripleader came back on: "No, that's left around the hole. Middle after that. Then right of center."

I couldn't read emotion in either voice. Tripleader continued describing chess moves, and I was looking out of our cove at a maelstrom river, thinking if it looks anything like this downstream, none of what he was saying could possibly make sense. I felt as if we were totally, irrevocably on our own. Signing off, we let him know we were coming.

Doc shouted with boat captain's authority, "Everyone got their lines? Smiles, everyone. Vomit now if you need to."

We checked each other's life vests. Dry suits were cinched, helmets cranked back and forth to make sure they were tight. We decided our order—who was first, second, third, fourth—and untied from shore, peeling away one by one. I gave Old Pola thumbs-up as he mounted one of the catarafts. He nodded and looked down, finding straps to hold on to. Each raft whirled back into the current. I clutched a fat yellow tube with my thighs, one foot dragging in the water, pushing us off the rock. We rushed around a few car-sized boulders and winged out of the eddy.

The first smack threw me to the slick, squeaky floor, and I scrambled back up as Berkeley Woman shouted for me to get my paddle in the water. Lifted to the top of a swell, I could see for just an instant as rafts ahead of us disappeared and burst back up through mud-brown explosions and haystack waves. I saw bright undersides and oar blades pointing at the crack of the sky. Water seemed to be going everywhere, and Berkeley Woman was shouting, "Draw, left, back, right!" She spun us, broadsided a wave, struggled straight. Rafts were starting to catch up with each other again, currents going everywhere. There was no downstream, no upstream. What the hell was Tripleader talking about?

We had opted to be last in line so Berkeley Woman could see

what was going on in front of her, but we were swinging ahead of other rafts now, caught up in a current that ended in a swaybacked plunge before a wave. I heard a voice shouting beside me. It was Doc, which struck me as odd because Doc was not in my raft. He was yelling, "Hang on for your fucking lives!"

I looked up only for an instant. His raft had shot up the side of a wave nearing vertical, another few inches and it would flop over on top of us like a sandwich.

He was not my business. My business was my own paddle.

I looked down and sank the blade, and that instant we went underwater. Split seconds split. Sound clamped tight around me. We were beneath the surface, and all I heard was the garbled roar of a monster, air bubbles shunted down through the mud. When we blew out the other side, Berkeley Woman's voice broke through, yelling to dig. The next wave was already on us. What happened to Doc? I didn't have time to look back before we blunted our way through the next crest.

In the next possible eddy, we cut from the momentum, and rafts piled into one another. The two point kayaks were waiting for us. Boat captains smacked the tops of their helmets, giving Tripleader the sign of all good. My blood was about as cold as it gets, maybe preparing me for an unexpected swim.

The two kayaks turned back into the flow like small birds darting in and out of view.

I had to remind myself to pay attention and try to absorb this place. Why I had come was not just to live. I had a job. I was here to look planetary tectonics in the eye and understand what it means to live at one of the great collision points of our time. This is what tectonics wrought. It is the next cosmic leap for humanity, first discovering the world is not flat, then discovering the world beneath the surface is not flat, either. I cracked an ammo box, pulled out my journal, and scribbled as fast as I could about lone pines grown from cracks and pitched up on ledges. The gorge appeared to be praying, as if hands were trying to meet.

The top of Mount Everest, highest elevation in the world and about four hundred miles southeast of my position, is made up of limestone that was once a seabed. Marine fossils are found near the summit, heaved up to twenty-nine thousand feet and still going. The rock making up this gorge was coming up from an even greater depth, as Old Pola said, Asia's basement. I was a local man from a much gentler landscape looking down the throat of action far beyond me, the Tibetan Plateau rearing back, expressing its sheer earthness, saying clearly that this is what is boiling beneath our feet. What I wanted to hear from the radio was Tripleader laughing and telling us to come on down, the water's fine. I wanted him to say that the rapid ended right around the bend and from there out it was as smooth as the scales on a dragon's belly. I wanted time. Time to appreciate this place. Imagine floating downstream through this crushing gorge with the back of your head cradled in your hands as if you were lying on the floor of a cathedral peering up.

The radio remained silent.

For half an hour we waited as my hopes flagged. When Tripleader's voice finally crackled from the handheld unit, he sounded far away. He was still conferring with Burly Montana when he finally started by saying, "Listen." As if he were trying to sell us something.

"Listen, you're just going to have to read it and run," he said. He fired off a few tips, some lefts and rights, a large recirculation hole, and, "There's this big whaleback boulder. You'll see it. Don't hit it."

He finished up with "Then tee up for a big wave train and you're out."

"All right, you heard him," Doc shouted. "Read it and run."

There was no break, no breath. It was all water, and I couldn't tell upstream from down.

At one point, I was going into a down slope backward. I could see what was ahead only by the look in Berkeley Woman's eyes, which were frighteningly wide. She appeared like a Tibetan guardian statue standing at the gates of the underworld, eyes all pupil and white. I braced and passed through the underside of a wave back-first. Her voice broke through the other side. I didn't hear the words. She just wanted muscle. My bones hurt. My heart pounded as I sucked air. I gave up on the idea of my body. I gave her muscle as if this were my only moment. We crashed over the next wave.

Born.

Born again.

And again.

My boys. I will reach you.

For an instant I saw a raft behind us with no one on the oars. The boatman had been washed over. His hand stuck out of the current and grabbed onto a strap, which dragged him like a fish on a line. Oar handles flew wild and undirected. I lost sight of this unpiloted raft as we crashed up the wall of the next wave.

The hole that Tripleader warned about came up fast, and I remembered something he said about going left. Berkeley Woman remembered, too, setting up swiftly as we skimmed down its edge. It was a hollow place, a gap in the river, and I leaned out, peering into its recirculating bottom, the kind of place that would eat a raft. One of our oars hung uselessly into the space as we tilted, and Berkeley Woman pushed her body into the opposite oar, grabbing hard to the current, keeping us on top. We skated past and were thrown downstream.

The rapids ended in a long wave train. Champagne bubbles popped around us. Rafts spun, kayakers leaned over the brace of their paddles. It had been about five miles of nonstop class 5, the kind of water that shames you with a glance, but we were all here. One last boatman climbed out of the river to his oars and patted his helmet with an all-clear sign. Old Pola let go his grip. The gorge was finished, sky opening ahead. Berkeley Woman's body folded over her oars as she laughed in disbelief at what she'd just been able to do.

That night brought more rain. We set up a kitchen, sliced vegetables, threw in chicken thawed from an ice chest. Survivor Woman pulled out an iPod with speakers and played Counting Crows on shuffle. We danced arm in arm while raindrops pearled the surface of the river.

Tripleader was elated, practically leaping. "This is world-class white water," he said. "I mean, it is that close to being unrunnable, but we did it."

His smile beamed into the rain. "Where logic hasn't worked for us," he said, "it must be karma."

We pulled out Jose Cuervo we had purchased in Lhasa—bottles wrapped in duct tape so they wouldn't shatter in the rapids—and danced around camp getting drunk while yak herders watched with amazement from the opposite shore. Old Pola was talking about the recent appearance of limestone, mapping in his head the undulating contact between Paleozoic basement rock and the bottom of this ancient gray fossilized seafloor that had been raised up over millions of years. Berkeley Woman and Doc were singing. Meanwhile, Outward Bound crouched in the mud with the Tibetan and Chinese crew members, keeping their cigarettes lit under their hat brims. They looked downstream at the dim limestone gates of the next chasm. Applaud ourselves as we might, we still did not know what was ahead.

As we floated down the river, mountains slowly turned their faces toward and away from us. We passed through moon-gray bedrock of solid limestone, a fossilized seafloor, what must have existed between Asia and oncoming India fifty or sixty million years ago. If you had stood in this place not long after the extinction of the dinosaurs, there would have been no mountains. Instead, a sea would have stretched around you, a particular world that could never last. At least thirty million years of mountains have since climbed each other's backs, former landmass and ocean bottom staggering upward, replacing that former sea, sending its floor to the sky.

We were getting good at this, following the breathing earth as landforms opened and closed around us. The river tightened and unfurled as it passed through this landscape of slow-motion calamity. Hey-diddle-diddle right down the middle, we slipped through rapids and waves like an arrow shot through the center of an apple.

We kept running into snags, big rapids, various minor setbacks, but nothing actually went wrong. Not terribly wrong. On sharp rocks in the middle of a rapid, we blew out a tube, rubber blubbering in the water as air escaped. As the raft half sank, I drove a hand pump to keep us afloat, working up and down on the deck through a tangle of white water, while Outward Bound rowed. Patched within a couple of hours, the raft was good to go. The next day downstream, the banana raft flipped with the Chinese man on the oars. It was a textbook flip. Throw lines shot out as if from cannons, swimmers pulled in. Tripleader and Burly Montana, aimed their kayak noses and bulldozed the capsized raft a hundred feet across the river, planting it in an eddy; they jumped out and performed a flawless two-person rope flip. Within fifteen minutes, the raft was back in service, no serious injuries other than some blood and a sprain, and not a single piece of gear lost.

Our last night, camp was on a cobbled bar. The river had come down far enough to give us half a ballpark's worth of space. According to the Soviet maps, we were now five miles from the paved bridge we'd be using as takeout. Our caravan of Land Cruisers and the supply truck would be there to meet us. We had originally planned two weeks to run the Gyalmo Ngulchu. The flood put us here in five days.

Skies were clear, air crisp with autumn. As the sun set, we bathed naked in cold water. It was our first cleanup in weeks, met with hoots and hollers as we jumped into a slow and chilly eddy. Heads plunked under came up gasping, our bodies finally free from moist dry suits. Old Pola kept baggy white underwear on for discretion.

That night, we gathered driftwood for a fire on the freshly exposed beach. Nobody had anything to say, staring into the coals as a log cracked in half. Sparks flew upward. They became many cool blue stars.

In the morning the fire was reduced to wisps of smoke. A bean can was used to heat some breakfast water in the coals. The river was down far enough to expose ranges of cobbles, hundreds of gray and black palm-sized rocks slipped one into the next. Their matrix was armored against the river's flow, rocks arranged like fish scales, one tucked into the next. Stability. Stasis even. The Buddhists were wrong, I thought. It isn't just about transience. If it were all pure, seamless blockbuster action hell-bent for entropy—pure transience—there would never be a patch of ground on which to live.

One of the cobbles caught my eye. I bent down, slid it out of place. On its face were worn etchings of Tibetan lettering. Blended in with all the other unmarked rocks, it was a prayer stone that had washed downstream. The river's actually littered with these things, shoreline shrines spilling in for centuries.

I weighed it. Its scripted font was wearing out but still there, fugitive, enduring, as if I were holding the very earth in my hand.

Hang out with geologists and other earth scientists for long, and they start bending your perception of time, like light passing around some massive object.

Travel through the topography of the earth, and the same happens.

I once went fossil hunting back home with Old Pola. We were walking desert hills in western Colorado, picking through fifty-million-year-old clams caked into layers of eroded badlands. Here we found evidence of a former earth, a local world ended long ago.

"Why are clams just in one thin layer?" I asked.

"Something happened," Old Pola said.

"Something?"

"An extinction or a mass bloom of mollusks, some kind of remarkable event."

In my hand I looked at a shell I'd dug up, a crooked little cup that once held a silvery bit of life. I, too, was made of bone, and not even as hard as a clamshell. I looked up from a crouch at Old Pola and said, "What will we look like in fifty million years?"

"What will *we* look like?" he asked, making sure he knew the scale I was talking about.

"This moment in human history. Us. Human civilization."

"Oh, hell, Craig."

That was something he often said when I asked such questions. "It'll be different," he said.

Old Pola came down beside me scraping a patch of gray-stone clams out of the ground. To answer my question, he thought about the unprecedented volume of carbon we've liberated, which would at least leave a dark blast layer similar to the K-T boundary that marks the end of the dinosaurs. He thought of our cities and freeways, our countless soda bottles, and what would happen to it all. He spread thumb and forefinger as if describing the size of an inchworm.

"We'd leave a layer about that thick in the side of some rock face," he said.

According to the Soviet maps, the last five miles didn't involve much topography. We expected mountains to step back like a curtain for a glorious ending. The maps were wrong, however. A seventeen-thousand-foot mountain of silver-gray limestone stood in the way, an ancient seabed shoved up into the sky. The mountain looked as if it had been opened by an ax blow, and we passed below a single wall about four thousand feet tall. Water calmed below it. Oars stopped moving. There was not even a wind as the current hurried us into spirals and turns. We gawked straight up the highest sheer slope. It looked as if you could stand on the mountain's remaining summit and pitch a rock all the way down to the river. In the face of this immensity, any hope for personal distinction we might have nursed this far down the river was finally squelched.

Most earth scientists I have spoken with talk of the immediate future as if it were nothing. They fall back on the idea that millions and even billions of years lie before us, and humans cannot possibly destroy the earth. We are too young, too quick, too ephemeral to really count. They say civilization is all but gone. Our species is drowned out. Relax, be happy, life will continue. For as much solace as I take in the long view, I can never escape the ravishing authority of the present, the small ticking of minutes and even seconds. It is the moments of being alive that most hold our attention, while those moments are part of a much larger framework of millions of years. It all becomes inseparable.

The next rapid caught us by surprise. If you keep them numbered, this was the thirteenth big one, and the river disappeared over an edge, roaring down into something unseen. A piece of the mountain had come off. Boulders had slid down, crossed the river, and gone up the other side.

Throwing themselves into their oars, pulling against the current, boat captains tried to separate rafts, everyone shaken out of personal reveries back to the reality of this river. There was no room to scout, hardly enough time to pause or think. We couldn't see over the edge, the rapid invisible. Shots of water appeared briefly and randomly from over the water horizon, blowing off in odd directions and with timing you'd rather not experience in a rapid. Signs of chaos.

Tripleader looked back from his kayak out on point and signaled everyone to run just right of what looked like a ceramic lip dipping over the edge.

He straightened, plowed his blades hard, and was gone.

The rest of the kayaks swept down after him. Then rafts, one by one. I was at the tail in a cataraft with Outward Bound on the oars along with the Tibetan crew member Chang Dak. Old Pola clung to the next raft ahead. He looked like a barnacle latched onto straps, one foot tucked firmly in a crotch of the cataraft tube, as Doc shouted at him to hold on for his ever-loving life, and over they went.

Outward Bound was up on his toes, trying to see what was happening. He dropped fast and said to Chang Dak, "Double up with me; we're going to need some power. Craig, up on the bow."

Chang Dak jumped to face Outward Bound, oar handles double fisted between the two of them as they multiplied their efforts on each stroke. They were going to try to power us through. Through what, I didn't know. When all else fails, get momentum and hope for the best. I landed on the point of the front-right tube as we slid down the tongue, picking up speed. For that second I caught sight of a mess of rafts half flipping in what looked like a pit of exploding glass. It wasn't a long rapid, just one sockdolager of a drop. If we'd been prepared, had it been scouted, it would be different.

As we raced toward the first ten-foot wave, Outward Bound shouted to me, "I'm going to stick your nose in it. You'll break through for us. Just throw your weight."

He held back on the right oar. We turned just slightly so that I came under the wave's rooster crest, which was collapsing beneath its own weight, a mass of water you couldn't even drive a car through. We went straight into it.

Certain Tibetan Buddhists have a name for this kind of moment, *dzogchen,* an awareness of the perfect self that exists in all things, a state of being that cannot be described. Masters meditate for lifetimes to achieve and sustain this state of being. Mine would not be maintained. It was there for only an instant.

I sank into the wave and breached into the air on the other side. My face split into a thousand pearls of flying water. Just about to breathe, I noticed a second wave, an even bigger one, on my right. The sucker punch nearly threw me as I found myself grabbing the raft tube like a monkey humping a log. *Dzogchen* all the way.

"Push, push, push!" Outward Bound bellowed with every stroke as waves struck from all sides.

It was a quick rapid. We were free and swirling on downstream foam. Each raft was upright. Kayaks bobbed. Water poured from under our helmets as if we had just climbed out of a washing machine. Smiles, slowly. Then laughter.

"Jesus Christ," shouted Chang Dak, a devout Buddhist.

Tripleader swiveled his kayak to survey his group. His face was framed in wonder. Everyone had come through. A few months later, his dad would find out that he went ahead with this venture even at flood stage, and he would get a good talking-to.

We passed through the edge of the mountain, and the next bend opened onto a bridge gleaming in the sun. We'd seen only a few other bridges along the way, each rickety and bunted with prayer flags, just wide enough to get a horse across. None looked like this, spotless and communist gray, standing like a dam. This would wash away, too. How long, really, could you maintain a bridge in a place like this? I gave it a century at most, and after that, ruins traveling downstream. Pieces of concrete would join wayward prayer stones eventually heading to the sea, where they would sink and the land below them would eventually ride under Asia and dive into the glowing mantle. That, to me, was more exciting than the idea of us leaving an inch of ourselves in a rock outcrop. This way, we returned, straight back to the source on a planet that never stays still.

Three police officers appeared, running out along the concrete rail, their blue suits and perky hats sharp against a clear sky. They had sent word up to villages to ask if we'd been seen, but no one had reported back. With the flood, they had been expecting the worst, boats and bodies. Seeing us alive, they seemed elated, stopping to wave down at us, the only thing to happen out here at this lone outpost in a long time.

Below the officers' small, round faces, just out of their view, was a thin concrete lip spanning the entire bridge where I noticed hundreds of prayer stones balanced on edge. Pilgrims had leaned over to place them on the lip, arching them across the river where only the water and the wind it carries could see them. The officers could not. As we came closer, I noticed that each stone was inscribed with the same letters and images the monks had drawn on us, prayers for compassion and light. Maybe it's no coincidence that this is seen as such a holy landscape, a place you come to send off prayers. It is one of the earth's pounding hearts, destroyer, creator.

I looked back as we approached the bridge. Walls of mountains overlapped, hiding the course of the river. We'd floated through a tectonic accordion, the very reason this planet tilts in its many different ways, from its earthquake-hammered axis to long-term climate changes. Because we have mountains, we are alive.

8

Cataclysm Strikes

Meteors are not needed less than mountains.

—Robinson Jeffers

HAWAI'I

BARNACLED WITH CINDER cones and hemmed in by sun-blazed clouds as far as the eye could see, the volcano Mauna Kea looked like the shell of an old turtle swimming through its own creation story. Its many-pointed red-black summit stood 13,796 feet above sea level, a massive, barren crust raised into a chilled upper atmospheric wind blowing across the Pacific.

I hurried through jumbled, hardened lava as if with some urgent message to whisper in the turtle's ear.

A metal cup banged on the outside of my pack as I jumped from boulder to boulder, my trekking poles click-clacking over cracks. The wind roared across bare rock, whistling at sharp crags. It was dry. It made your lips crack. Down below, there was no gap or window in the clouds, no sign of ocean or earth, only a cotton-white firmament stretched infinitely in all directions. The island of Hawai'i, which should have been visible down there, was completely obscured as if the only thing that existed was this shield of a sleeping volcano capped beneath the blue-domed mirror of sky. It felt like the last thing in the world. Or the first.

I stepped wide across fissures that once smoked and hissed like hellish dragons, now quiescent but for wind. The place was as motionless and freeze-dried as the surface of Mars. It was late afternoon, and Mauna Kea's shadow stretched at least a hundred miles over the clouds, nearly touching the horizon. Soon the sun would set, and I didn't want to be late. It was New Year's Eve, and I had my own little party planned. Exactly at the moment of sunset, the full moon would rise on the opposite side of the sky, a perfect balance that happens once every month. Not only that, it would be a blue moon, the second full moon of the month. In this rarefied air, I figured this had to be one of the best seats on earth tonight. I wanted to be in the right place at the right time to watch it happen.

By most reckoning, Mauna Kea is the tallest mountain in the world—thirty-three thousand feet when measured from its seafloor base, four thousand feet taller than Everest. Its entire mass is made of lava roped and dumped on itself for about one million years.* The growing weight of the Big Island, Hawai'i, presses down on the oceanic crust, giving Mauna Kea an actual height (measured from the bottom of its convex base) of about fifty thousand feet. Near its top, crusty snow gathered in the shadows of cinder cones under the full-bodied roar of wind. I found the proper viewing spot on the east slope of a ruptured cone. There was nothing alive that I could see. With a daily temperature range between 116 and 25 degrees Fahrenheit with practically no soil, this is not a place where life can make much of a stand. As much as it felt like a creation story, it also felt like annihilation. This volcano was the first solid thing to form and the last thing left.

Mauna Kea had not erupted in forty-six hundred years—unlike other volcanoes on the island surrounding it, flowing almost constantly in the historic record. The violence was frozen in place, magnificent fountains of lava showered down big, steep slopes where boulders the size of cars had fallen on a once hot and gassing surface. When boulders hadn't been raining down, glowing cannonballs had streaked from the sky, half-molten projectiles with tails tapered, and hardened as they fell.

Gloves on, hood secured, I unloaded at a comfortable place to sit, pulling out an assortment of gear. Fitting into layers and tucking in

---

*The Hawaiian Islands formed from eruptive seamounts building up into mounds of hydroexplosive rock as lava hit water and air at the same time. Lava born into just air poured down the sides into the ocean, layer by layer covering the old seamount in broken-glass rock and pillow lavas miles thick.

everything else I had around me, I snugged into boulders facing due east, wind snapping over my head. For tonight's party, I brought out a tub of creamed-rice pudding and a sack of fried drumsticks that I had picked up in Hilo. No party favors, though. No fireworks or sparklers. I couldn't get anything lit up here if my life depended on it. As if settling in at a drive-in theater waiting for the show to start, I spooned up rice pudding and scanned the horizon. If anyone else was on the mountain this evening, it would have been the few astronomers and technicians stationed at a cluster of silver-hooded telescopes nearer the summit three miles behind me. I wondered if they were cracking open their chamber doors, their faces braced into the wind as they checked the sky to see if it was time yet.

Almost.

The shadow of the earth moved across clouds. Evening grew. A halo of light began to rise in the distance. After several minutes, a bead of molten glass appeared on the horizon. I set the spoon back down.

As the moon came up, it poured over itself in obese folds. A hot-pink shape lifted through waves of atmosphere. Soon, the squashed oblong pulled free, parturition headfirst. The injuries to its face were visible even before it cleared the horizon. Lifting before me was a museum of collisions and ancient impact-related eruptions. A quarter the diameter of earth, the moon is visibly decorated with craters, ejecta rays, and dark, iron-heavy lava flows four billion years old where lunar guts had been squeezed out by hard blows. You see in these scars the grotesque man in the moon or the fleet Maya rabbit.

Usually, a full moon would be too washed-out to view straight on, its finer features turned to atmospheric glare, but this thin, dry air was like looking through window glass. The nickel-bright sphere lifted to eye level, and it looked like an actual *moon*, a huge pock-marked rock floating inexplicably above the earth's surface. I swear I could have reached out with a pole and touched it.

I pulled binoculars, an old pair of metal Leica 8 x 20s. The moon exploded. I was instantly on its surface, moving freely across the incandescent blast of the Tycho crater, then up to the serene gray crater fields of the far north. I circled the bright rim of the Copernicus crater fifty-seven miles across, landing on a wink of shadow in the center, a cluster of hillocks, rocky splash back of an impact. Feathery ejecta rays spread five hundred miles in all directions from Copernicus, remnants of a strike estimated to have occurred 800 million years ago when an object the size of several shopping malls impaled the lunar surface—something that would have easily penetrated the earth's atmosphere and touched down at over a hundred miles a second if it had drifted our way.

The so-called dark side of the moon, which we never get to see from here, is far more beaten, with twice as many craters, like rain-pocked mud. It is perhaps fortunate we are not looking at that terribly cratered side all the time, or we might be throwing our hands over our heads whenever it rises. The backside of the moon is our shield, giving some approximation of what did not hit us; its surface what is called saturation equilibrium, where new craters are so numerous they absolutely cover any evidence of older ones.

The majority of the moon's larger features, the terrain beneath many of the craters, came from an event known as the Late Heavy Bombardment, an asteroid free-for-all when objects perhaps hundreds of miles in diameter careened in all directions, some hitting the earth. The asteroid belt (now found between Mars and Jupiter) was disrupted by a gravitational swing between the outer gas-giant planets, which flung each other into new orbits. Chaos ensued, asteroids flying every which way pummeling the inner rocky planets, earth included, and leaving behind a moon looking like a face soundly beaten in a fight. That was around four billion years ago. This bombardment went on for at least twenty million years, reliquifying large portions of our planet's surface into molten rock in what is widely accepted as the beginning of time on earth.

The moon is the better storyteller for this event. Our ancient craters are smoothed over by erosion and tectonic motion. With no erosion, no wind, and no liquid water on the moon, craters can remain perfectly visible for billions of years, an orbiting catalog of impacts. Lunar soil holds precise dates of these impacts in tiny glass spherules that melted out of vaporized rock every time a big one hit.* Extracted from powdery soil brought back by Apollo astronauts, the dates of these spherules show patterns when impacts came in waves, not only the Late Heavy Bombardment, but a second assault of asteroids shortly after. There was a lull for a couple billion years followed by another spike about 400 million years ago, right around the time life really started going on earth, as if jump-started.

The Late Heavy Bombardment is what everything else is measured against, an apocalypse unmatched in earth's recorded history. Prior to the bombardment, Mars probably had more of an atmosphere and hydrosphere than at present, and also appears to have had a magnetic field much like the earth. One hypothesis is that impacts on Mars that coincided with these terrestrial and lunar bombardments turned Mars's atmosphere and hydrosphere into a steamy greenhouse. This may have resulted in evidence of former lakes, rivers, and streams found on the Martian surface. As the bombardment continued, most of that atmosphere and liquid water may have been blown into space, leaving the planet as we see it today, its magnetic field destroyed and any remaining water frozen at or below the surface. If there was life on earth prior to this event, there is no record of it. The oldest terrestrial rocks we know of formed at the tail end of the Late Heavy Bombardment—dating to about 3.8 billion years and found on the coast of Hudson Bay in northern Quebec, where North America was first to emerge from a semi-molten earth. At the time, the earth was enwreathed in lifeless, volatile vapors. This was the beginning genesis, the first recognizable sea and atmosphere formed in a burning, primordial mist. The moon would have been much closer back then, its bright circle dominating the sky. Had I the ability, I would have sat right here watching it unfold, a fly on the wall at the time of creation, or the time of destruction, hard to tell the difference. I would have pulled my hood, shielding my face from

---

* Bolide impacts are generally forceful enough that rock vaporizes and then re-forms as glass droplets smaller than pollen grains. These spherules, as they are called, fall back to the ground, where in any given gram of lunar soil you can find tiny glass beads from a hundred or more impacts.

the wind as it whistled through zippers and stays on my outer jacket, looking up at a freshly cratered moon, the world yet to be.

Wind sang across barbs of volcanic rock.

Milky light made shadows out of boulders.

I wagged a fried chicken leg at the moon and announced, "Happy New Year!"

The end that I was looking for this time was the surface of the earth wiped clean, not the symbolic, geographic cleansing of tectonics, but total, instant annihilation. By sunrise, the full moon hung upside down on the other side of the sky. Sunlight sheeted across clouds that still lay far below, and it struck me where I sat against the same east-facing boulder.

I had slept at the viewing spot, my bag curled like a worm in the boulders. I crawled out of my cocoon. The wind had not paused. As I strolled into the sterility of this high volcano, the moon set among reddish summits and the silver-white domes of observatories. Cinder cones were the size of Druid mounds, some pocketing crusts of gritty, melted and refrozen snow that looked like fallen mirrors, like pieces of sky gathered on dark earth. Jets of dust shot out from under my slow steps as I looked toward the clouds below, flat as they were last night, only now pierced by a pillar of smoke. The smoke emanated from the active east rift zone, a four-mile-long crack coming off a dike-fed crater called Pu'u 'O'o, the source of the most land-consuming eruptions on the island since about 1983. It was more than ten thousand feet down from me, about thirty-five miles out on the lower flanks of Mauna Loa, Mauna Kea's fiery and iras-cible sister to the south.

Volcanoes and bolide impacts fall into the same category in my mind—random, explosive, devastating. The biggest ones are like glo-bal battering rams. In the astronomical corner, there is the Chicxu-lub impact (pronounced "cheek-shew-loob"), the asteroid strike over what is now the Yucatán Peninsula, believed to have taken out the dinosaurs. In the plutonic corner, you see a swarm of super-volcano eruptions from Colorado to the Mexican Sierra Madre around 28 million years ago, leaving at least one caldera from a single explosion more than seventy miles across, five thousand cubic kilometers of matter blown out all at once. Recent history is peppered with smaller versions that are happening more or less all the time. In 1815, the Tambora volcano erupted in Indonesia, the largest eruption known in the past ten thousand years. Coinciding with a historic low in solar activity, it is believed to have sent the world into a debilitating volcanic winter, known as the Year Without a Summer. Food stocks crashed worldwide as crops failed in Europe and North America in what has been called the last great subsistence crisis in the Western world. Considering that forms of agriculture we rely on for substance have not changed terribly since the nineteenth century other than becoming industrial in scale, you might want to consider what a historically large eruption could do. Usual climate change models do not take these kinds of random, unthinkable events into account. Why would you bother plugging in their numbers? It's easier just to pray and hope you're living during one of the gaps between major events. But there may be more to survival than that.

As the sun peeled back the night's cold, I pulled off layer after layer, lashing them to the outside of my pack, and I became a ragman walking across what looked like Mars. It sounded as if I were stepping on potato chips. The ground was made of rubbled scoriae and lava-blasted malformations; this was a glimpse into what the planet can put out. My steps slowed, mouth choppy in this dry wind that was no longer cold. I crossed the red-grit flank of a cone, and the slope ahead of me rippled with heat. A film of a faint mirage wept into basins of granular exposures.

I stopped in a bed of warm lava dust winnowed as fine as cat litter. Rough little bits, I picked one up and rolled its ugly head between my fingers, thinking this would make a dart of light if it fell in the night sky, the size of most meteorites we see. It was volcanic, a creation from underground. I pulled off boots and stepped barefoot among these warm, cinnamon-colored cinders. My feet were still cold from the night, and quickly they warmed. Taking slow steps on soft skin, I thought this is what terrestrial life was made for. This is the beginning of habitat, something resembling soft. But how long does it last? One bombardment is followed by another. The slate gets wiped clean again and again. When and how will the slate be wiped clean again?

I crouched at a boulder, finding in its wind shade a row of little spleenwort ferns, *Asplenium trichomanes,* grown right out of a crack in the rock. If this is how clean the slate is left, I'd take it. I asked Edward O. Wilson how far is too far when it comes to extinction. "You know the answer to that," Wilson had said with a smile. "We can go all the way to microbes. The question is, how far do you want to take it?"

The ferns were finger length and as thin as parsley, tips brown from wind and freeze. Small, rough plants, they came before the more common vascular plants we know today, the oak trees and red roses. From this fern sprang the world.

The top of Mauna Kea would have been a dead ringer for Mars except every fifty or a hundred yards you'd spot a golden stub of xeric grass or tiny licks of ferns huddled in boulder seams. I was looking for a littler sharper taste of cataclysm, though. I was looking for a lifeless earth, one taken back to its barest beginning, so I switched volcanoes. I went from the frozen demeanor of Mauna Kea, goddess of wind, ice, and clouds, over to the fresh-lava surfaces of Mauna Loa, the most constantly active volcano on earth. It was a short hop, both volcanoes touching shoulders on the same island. That's when JT showed up.

On his way from South America to Alaska—somewhere between stops in New York and Colorado—JT was dropping in for a mid-Pacific visit. A friend from back home, he had ʻohana on another one of the islands, a Kauai family who had taken him in. JT had families all over the world. He worked as a freelance photographer in the landscape-and-culture department, hired by the U.S. Geological Survey to visually record Alaskan glaciers, frequently heading to Cambodia to cover dam issues on the Mekong.

I picked up JT at the airport. He plugged his iPod into our rental, bluegrass from home. He moved his seat back, planted his boot soles on the dashboard. "Take me there," he said.

JT was a medium-stature man in his early forties, with olive skin and a brave, Roman face. He had a house in Colorado where he lived with his dog, in the summer irrigating a hay field, in the winter keeping the woodstove stocked. He had a bit more stamina than Angus in the cornfields; then again, the rest of this trip had the potential to get a hell of a lot worse than corn. Industrial agriculture may be a fine analogue for mass extinctions, but this is a picture of total destruction, a planet rendered to metallic oblivion.

Before JT arrived, I had been hunting live volcanic breakouts to see some fresh coverage. I'd drive around back roads, standing on the roof of my rental with binoculars trying to see a glow. I was here with my family, Regan and our two little boys, and we'd catch glimpses of lava opening and closing, spilling into the sea. We had run into a lava junkie living at the edge of the most recent flow. A retired carpenter from the Midwest, he lived with his taciturn wife and a few mongrel island dogs in the last house in line for incineration. They had us in to show us maps and photographs, Lava Junkie saying living here was unlike anything he had ever known, so strange and beautiful that he found himself addicted to the volcano's power. He was fascinated by the devastation going on out his kitchen window, where every morning he looked upon a wasteland emitting pillars of smoke. The rest of the subdivision west of them had been covered by molten rock flowing in stages over the past several months. Charred and fallen trees lay like elephant skeletons across a quilted surface of black and silvery lava. A street sign stood out of quilted basalt, and near it was a burned-out, flash-rusted car whose chassis must have lit like a candle wick as it was carried along, its seats burned down to metal springs.

On a small kitchen table Lava Junkie spread aerial photographs and maps that he had dotted himself showing different dates and flows over the last few years. He showed sleeves of photographs he had taken of lava rivers moving through trees and houses, rolling over paved streets.

"You've seen a lot disappear," I said.

"No," he said. "I've seen a lot created."

One morning around three thirty, Regan and I pulled the boys from their tent, dressed them in coats and boots, and tottered them out toward a fresh outbreak that had begun exploding into the sea. It was a brilliant predawn maneuver, the boys' faces pumpkin lit with glee and lava glow. (For any anxious parents, note that we remained a safe half mile back in a steady upwind the entire time, our exit route duly prepared.) If I wanted to get closer and actually walk across the smoldering middle of this shield, I had to ditch the family. They flew back to the mainland a couple days before I headed for Mauna Kea to take a break from all the blackness. Then JT showed up, someone willing to go out and do some half-stupid thing with me in search of the second coming of the Late Heavy Bombardment. Our plan was to backpack across the active lava shield and see what such a destroyed planet might look like.

Lava Junkie offered me and JT a spot to camp on his property while we waited for the next outbreak. The lava had stopped flowing on the surface and was building underneath, pushing up on the hard volcanic shield causing it to shift and bulge. "It's going to be soon," Lava Junkie said.

The plan with JT was to start backpacking across it as soon as the first glow appeared. We wanted it fresh.

Every time a big space rock flies silently past our globe, we let go a collective shudder as if a shadow just passed our tender earth. We are a sitting duck out here, this blue globe spinning at a thousand miles per hour while traveling around the sun at sixty-seven thousand miles per hour, a gravitational long shot for anything that flies by, but a naked target nonetheless. The moon might take the hit instead—which would introduce its own travesties as chunks of lunar ejecta come flying at us. Our atmosphere tends to sizzle most smaller objects that come our way, but the big ones punch straight through.

On November 8, 2011, a roughly spherical rock measuring about 1,300 feet across passed between the earth and the moon. Named 2005 YU55, this asteroid came within 201,000 miles of us, a distance that gave little cause for alarm. It was a subject of curiosity, however, a quiet reminder of what is wandering around the inner solar system. The Keck Observatory atop Mauna Kea ran a live Webcast of the asteroid's progress, catching a parting shot in which it looked like a bright, sunlit seed flying away.*

To put a possible 2005 YU55 impact in perspective, go to Barringer crater, a tourist stop just off old Route 66 in northeast Arizona where an asteroid estimated to have been a mere hundred feet across hit during the Ice Age. Out of a flat and spare desert plain, the rim of Barringer rises 150 feet, as if a stone had been dropped into cake batter. The impact left what is now a vacant crater measuring 570 feet deep and 4,000 feet across. At least half the asteroid's mass probably made it all the way to the surface. Fragments have been found that survived the explosion, the largest piece recovered inside the crater weighing fourteen hundred pounds.

If a bolide is large enough to make it through the wool of the atmosphere, it will slam a hole right through the crust. It can easily be miles wide. Massive volumes of exploded rock will eject back into the air and even into space. Most of that comes raining back down all over the planet, some pieces enter decaying orbits where like old space stations they circle for years before eventually falling back. So it doesn't end with a first impact. After that comes skies of fire, return pieces igniting in the atmosphere.

---

* The next notable asteroid predicted to come near is known as 99942 Apophis, a space rock about 885 feet across that will approach in 2029 and again in 2036. Although there are significant gravitational margins when looking that far ahead in time—near-earth space objects can be erratic as they are pulled and sling-shot around gravitational wells of other planets—it appears that this one will miss us, predicted to pass at least 29,000 miles away from hitting us.

If the impact is big enough, waves of pressure and heat move through the inside of the earth like a kick to a water balloon. One theory is that a plug blows out the other side where you find future volcanic hot spots, such as what feeds Hawai'i, (or in that spot you might see a breach of flood basalts like the lavas covering most of the Pacific Northwest thousands of feet deep). In defense of this controversial antipodal-volcanism theory, of the forty-seven recognized hot spots worldwide, twenty-two are opposite signs of impacts (shocked quartz, tsunami marks a hundred feet high, and spherules of melted glass). Models of a forty-thousand-megaton surface explosion from an incoming extraterrestrial object concluded that extinction-causing flood basalts in the past could have easily been triggered by a large strike. It may be no coincidence that a few hundred thousand years after the dinosaur-killing Chicxulub impact, the Deccan Traps of India erupted, one of the largest flood basalts in history, not quite on the exact antipodal, but at least on the other side of the planet.*

We live on this thin skin between below and above, volcanoes rumbling and space busy with objects drawn to the gravitational bull's-eye of our planet. A one-year study out of the University of Adelaide in Australia picked out the "echoes" of objects the size of a dust grain on up to baseballs and bedrooms as they ionized the upper atmosphere on impact. The study detected 350,000 entries into the atmosphere in that year from one observation site alone. Some estimates go up to billions of objects four ounces or more igniting around the world every year, as if we were steaming through faint clouds of rocks and dust. Over time, though, we meet larger objects. The Manicouagan crater looks like one giant eye in the face of Quebec. It was formed about 215 million years ago when something

---

* Chicxulub was probably not the only major asteroid impact in that period. What appears to be a much larger crater was found on the seafloor off the coast of India. Estimated to have formed a few hundred thousand years after Chicxulub, this impact may have originally put a three-hundred-mile-wide hole in the earth's surface. A coauthor of the 2009 report on the Indian crater, Sankar Chatterjee, a paleontologist at Texas Tech University in Lubbock, has called this the Shiva crater, referring to the Hindu god of creation and destruction.

three miles wide touched down, leaving an outermost ring about sixty miles across and an inner ring forty miles wide now rimmed with a deep-blue lake.

More recently is the famed Chicxulub crater, thought to be the end of the dinosaurs when it struck sixty-five million years ago. Discovered in the 1970s, this is the one impact that has most enchanted human imagination and stirred controversy. Its remnants appear to be invisible to the naked eye. Best defined by concentric circles of gravitational anomalies where rock was once shattered and compacted, this crater straddles the Gulf of Mexico and the Yucatán Peninsula and is more than a hundred miles across. It was probably caused by an asteroid at least six miles wide, a mountain of rock and ore hurtling through space, and when it hit, it left shorelines around the world now fossilized with evidence of tsunamis at least a hundred feet tall, its explosion and aftermath throwing the world's biota, if not into a tailspin, then into instant global annihilation.

While our atmosphere weeds out most of what falls toward us, it has virtually no effect in slowing a Chicxulub-sized object (estimated at one every hundred million years). The closest contender at the moment, hardly a Chicxulub, is a four-thousand-foot-wide asteroid that currently has a one-in-three-hundred chance of hitting the earth in the year 2880.

This is all that can be known, chains of probabilities however slight. Seven thousand notable asteroids have been documented within range of the earth, where at least one asteroid is actually tagging constantly along behind us like a faithful dog. This is the context in which we live, a rain of dots and dashes in the sky intermixed with periodic brighter flashes, some leaving modest craters or at least brightly colored vapors in the atmosphere. Every hundred million years or so a world-ending splashdown happens, but chances of this occurring in any nearby lifetime are so low you really shouldn't trouble yourself other than to know that it does actually happen.

Meanwhile, about two hundred volcanoes are extremely active right now, several eruptions per day. So you'd be better off looking below than above if you're waiting for this kind of disaster. Steve

Sparks, a professor at the University of Bristol in the U.K. who studies volcanic hazards, estimates that we could see several Toba-sized super-volcanoes in a period of 100,000 years (Toba erupted about 73,000 years ago, believed to have nearly killed off the burgeoning human race). Sparks wrote,

> They occur more frequently than impacts of asteroids and comets of comparable potential for damage. Several of the largest volcanic eruptions of the last few hundred years (Tambora, 1815; Krakatau, 1883; Pinatubo, 1991) have caused major climatic anomalies in the two to three years after the eruption by creating a cloud of sulfuric acid droplets in the upper atmosphere. . . . Super-eruptions, however, are hundreds of times larger than these recent events and their global effects are likely to be much more severe. An area the size of North America or Europe could be devastated, and pronounced deterioration of global climate would be expected for a few years following the eruption. Such events could result in the ruin of world agriculture, severe disruption of food supplies, and mass starvation. The effects could be sufficiently severe to threaten the fabric of civilisation.

A super-eruption is what we talk about stewing beneath Yellowstone in Wyoming, an explosion unlike anything seen in human history waiting to happen. Such an eruption is defined as putting out more than a thousand cubic kilometers of ejecta. Mount St. Helens sent up about one cubic kilometer. The one that left the seventy-mile-across hole in Colorado, known as the La Garita eruption about twenty-eight million years ago, is what remains of the world's biggest known explosion. That caldera is now populated by forests, streams, and clear alpine lakes. But you can still see the sheer cliffs and uplifted, mountainous interior, the caldera itself so large it takes in the entire landscape. These exposed cliffs tell stories of whole mountains raining down, semi-molten masses in the millions upon millions of tons, incandescent ash landing all at once in a single layer five thousand feet thick. Billions would likely die if it happened now, not merely those in the immediate path of destruction, but those dependent on agriculture worldwide as temperatures plummet and weather becomes even more erratic. Thirty million years ago was a particularly trying time, and not from this explosion alone, but from volcanic chains blowing off throughout the North American Southwest from Oregon to the Sierra Madre, which may have been responsible for the Eocene epoch ending in icehouse conditions and Antarctica first becoming coated in glaciers.

Look inside the La Garita Caldera and you will see the eroded remains of Yellowstone-sized super-volcanoes that blew off after the big one, coming up inside of the original hole. These eruptions, each big enough to be a civilization crusher in itself, went on for a few million years. Surrounding the big caldera, you see ten thousand square miles of ragged volcanic remnants, the sawteeth and stony spindles of the San Juan Mountains, a landscape of numerous overlapping super-volcanoes plumbed with the throats of huge, Rainier-style stratovolcanoes that were active once upon a time. Pieces of bedrock in the area were uplifted eight thousand feet from their former positions.

What does all this do to life in the big picture? The answer is not clear. Volcanic activity during human history has often been shown to bring on cold spells resulting in failed agriculture and starvation, so we know it does something. But if you look at the record of bolide impacts and volcanism over the last sixty-five million years, these events don't seem to line up perfectly with extinctions or biotic turnovers. And throughout earth's history, both bolide impacts and major volcanic episodes occur much more frequently than mass extinctions, giving doubt to cause and effect. At the same time, some disruptions in the earth's crust go hand in hand with mass extinctions. It may simply be a matter of timing.

Dinosaur history, for example, is full of ups and downs where a destructive global event may have coincided with one of the downturns to wipe them out. We typically see dinosaurs as having a glorious 165 million years of thundering, unstoppable domination, with a sudden and mysterious end. In reality, this was not the case. As spotty as the fossil record is, it shows dinosaurs experiencing numerous turnovers and rebounds. Whole lines were wiped out and replaced, some dinosaurs barely squeaking through to rise to dominate in different forms. The last few million years of their existence in the fossil record shows a general decline in global health, species steadily fading from unknown causes, as if the earth were ripening for a mass extinction. Then came an "event," a deathblow that may have taken from hours to hundreds of thousands of years. The last dinosaurs vanished in a tumultuous period where failures may have been stacking on one another, leading to a final fall. Evidence of more than one major asteroid impact has been found on different parts of the globe around that time, again, within a few hundred thousand years of one another. There were also enormous flood-basalt volcanoes going off, forming what is now the Deccan Plateau of west-central India. These would have scrubbed the atmosphere with poisons and potentially climate-changing gases. Rather than being any one, unequivocal disaster, these may have simply been the straws breaking the camel's back. Bolide impacts and globally destructive eruptions may be happening all the time, but at certain moments they line up with intrinsic failures for a coup de grâce.

In that case, there would be better and worse times for the planet to see a super-eruption or an asteroid impact. Judging by our currently destabilizing climates and the decreasing health of global biodiversity, this could be one of the worst times.

The glow appeared one night. It was dusk, and we were standing with Lava Junkie in his backyard, if you can call fifteen-year-old regrowth from solid lava a backyard. He had planted some papaya trees, and ferns were growing up through fissures. The amber light was several miles away on the high flank of the east rift, a place where the lava shield had melted open. We traded binoculars back and forth, focusing on a vaporous crimson shimmer in the air. Out from underneath it appeared what looked like mandarin lanterns being carried across black fields. The lava was flowing, rivers coming out. It was our cue to go.

We began an hour before dawn. Hitching up our packs, we walked around barricades where an asphalt road ended abruptly where lava had recently poured across it. Beyond that, the world became an absolutely blank slate. Lava Junkie walked us out the first desolate mile and then stopped. "I've never gone any farther than this," he said.

He gave us a few simple warnings, saying the downside of the Pu'u 'O'o crater is rumored to be very difficult travel, the hardened lava breaking under you with nearly every step as if you were walking on skylights. He wished us good luck and turned back.

If I was ever looking for perfect oblivion, I had found it. The preexisting earth was completely and utterly wiped out. We avoided some places, imagining their skin too thin to hold weight. What did we know, though? We were strangers here.

"I mean, is this solid?" JT asked, tapping a trekking pole on what sounded like a barrel drum.

I had picked up a sturdy papaya-tree walking stick along the way. Some other visitor had left it during an earlier flow. Its tip was black from where that other person had irresistibly poked it through the belly skin of hot lava. I pounded the stick on the glazed-black floor beneath me. It sounded hollow, too.

"Is this one of those skylights that breaks through, and way down at the bottom you land in a river of lava?" JT asked.

"Maybe."

I knelt, put a hand on the rock. It was warm enough that I had to tap my fingers. The flow had come through three weeks before, according to what locals told us, but there were hot spots beneath, lava pushing up unseen.

"What do you think is underneath us . . . if we fell through?" I said.

JT tapped his pole again.

"White clouds, cold beer."

"And beautiful women," I added.

"They become more beautiful the farther you fall."

I tried not to let my mind wander too far down into that particular rabbit hole, the ground around us caving in as JT and I pitched to an untimely end. As the poet Mary Oliver wrote, "No one, / moreover, / wants to ponder it, / how it will be / to feel the blood cool, / shapeliness dissolve."

Only in this case, blood would boil like grease in a pan.

Our shapeliness would indeed dissolve.

The sun climbed higher, and there was no shade, at least no shade any animal would keep. There were holes and cracks out of the sun, but they were either substantially hotter or decorated with iron needles from where lava stretched apart and then hardened into medieval torture implements.

It was too hot to sweat. We went around exploded *tumuli*, bulges of rock that had slowly popped open into exploded blisters ten and twenty feet tall, pushed up by lava actively gathering below, lava we could not see. The air rippled furiously above them.

The world was made elemental, turned into glass and steel and the drifting smoke of sterility. One good six-mile-wide asteroid could make a landscape look like this. Or a live volcano. Glossy ropes and potbellied rolls spread into a geographic carnival, a dolloped, overflowed landscape turned to crust-crackled stone. In the rising sun, the exterior gave off a silvery sheen. This particular flow was seven days old. We crossed over whorls and protuberances where iron and silica had squeezed and poured, then hardened into place. JT made statues of himself, his camera turned one way and the other, zeroing in to photograph some small thing, a hellish yellow wormhole letting up steam where he knelt on warm rock.

If volcanoes change the world, asteroids do the same, only more rarely, on a much grander scale, and much more swiftly. Whichever analogue we were walking across, the deed was done. Nothing remained living. The Late Heavy Bombardment was long enough ago it is hard to scientifically see what it would have looked like, though I picture it like this. A little more clear is the Chicxulub impact sixty-five million years ago. Doug Robertson, a recently retired planetary geologist and bolide theorist from the University of Colorado, has developed a scenario. "When the dinosaurs were killed," Robertson says, "I believe an asteroid did it and in a matter of hours. By the end of that afternoon, dinosaurs were all dead." Robertson sees the impact of the Chicxulub asteroid, an object around six to nine miles across, splashing a mixture of vaporized and solid matter into the sky. Some went into space, while most came raining back down over the entire planet. In his model, skies would have basically turned to flame everywhere. "From anywhere on the planet, you would have seen the equivalent energy of several one-megaton hydrogen bombs going off above you." Robertson calculates that this initial superheating would have lasted up to an hour and covered the globe. "The sky was red-hot, it was the color of the broilers in your oven. Stick your hand in a broiler oven sometime, and tell me how long you want to leave it there." An hour, he figured, was enough to cook every standing dinosaur.

Now take that picture and apply it to Chicxulub-size asteroids hitting one after the next for tens of millions of years during the Late Heavy Bombardment. Could something have survived such complete obliteration? Would anything survive if it happened now, the earth pummeled sometimes with half-moon-sized objects?*

The obvious answer is no. Even arsenic-eating blue-green algae that live around volcanic vents, the heartiest life on earth, would die instantly upon touching molten rock. But could there be another way to survive a complete resurfacing of the planet? During an impact, pieces of the earth's crust are thrown upward. If the impact is big enough, this debris reaches escape velocity and flies into space, where it either keeps going into the solar system, falls back immediately, or enters a decaying orbit around the earth, returning piece by piece over the next five thousand years. In these orbiting rocks, there could be life. Microbial ecosystems have been found up to a mile

---

* Experiments have replicated conditions of impacts into water that chemically resembles early ocean. The resulting environment may not be as sterile as one might think. Mixtures of solid carbon, iron, nickel, water, and nitrogen have been experimentally shot together at high speed, and from them researchers recovered traces of organic molecules such as fatty acids and an amino acid created by the impact. The same has been found in high-explosive volcanic experiments, as if earth cataclysm has biomolecules written into its design.

deep within the earth's crust, and it is conceivable that the members of that ecosystem could end up blasted into space intact within their rock.

Say you pitch a good thousand-pound earth chunk into space during a bolide impact. Bugs and worms that weren't instantly vaporized are stripped away at escape velocity. If any invertebrates somehow made it that far, once in orbit they popped, sizzled, and froze. But what of microbes in cracks within the rock? Experiments on two bacteria—practically indestructible *Deinococcus radiodurans* and the common soil bacteria *Bacillus subtilis*—have proved that they both could survive impact conditions and a long stay in outer space. Falling back to earth over five thousand years, if they survived reentry, they might return in time to see hospitable conditions once again. That is a lot of ifs, but it is at least hypothetically possible to reseed the earth from a cataclysm of asteroids and meteorites.

As soon as the Late Heavy Bombardment ended 3.8 billion years ago, the earliest life springs into the fossil record as if it were already running. For such a quick appearance, life must have either evolved during the bombardment itself or existed here already. The picture that is beginning to resolve is one in which life did not first form 3.8 billion years ago, but came from an earlier time and somehow made it through the cataclysm. Llyd Wells, an astrobiologist at the Center for Astrobiology and Early Evolution at the University of Washington, writes, "If the evidence is correct, it suggests that early Earth life dodged a statistical bullet, emerged elsewhere than the planetary surface, or originated (perhaps rapidly) after the last sterilizing impact."

Whether life reseeded out of meteorites or somehow survived in situ, the conclusion is that an inhabitable earth may have predated the Late Heavy Bombardment. If life existed here before 3.8 billion years ago, a livable earth may have been here pre-bombardment.*

---

* It should also be noted that the Late Heavy Bombardment may not have been the wholesale resurfacing it was thought to be. The University of Colorado research associate Oleg Abramov reexamined lunar spherules that had produced cratering ages for the moon. From them he set up a computer model, plugging in estimated asteroid sizes, frequency, and distribution. By doing this, he gained a framework for what might have happened to the earth during the bombardment. His model shows a surface with large

Mark Harrison, a Carnegie fellow in geochemistry and director of the Institute of Geophysics and Planetary Physics at the University of California, Los Angeles, found geologic signs of a whole other world that once existed. His pioneering work on early-earth crystal zircons found that these crystals could be used as timepieces, and also as mineral signatures indicating environments they formed in. Looking at grains of zircons found in Australia, Harrison believes they date back 4.4 billion years. In them, he found minerals made in the presence of liquid water, which is not the picture we usually have at the beginning of our planet. Harrison believes he has found evidence in these zircons of continents, seas, and even the marginal processes of tectonics. The more common view he is challenging is that the early earth had a boiling, molten surface on which nothing could live. "This is a radical departure from conventional wisdom regarding the Hadean earth," Harrison said. "But these ancient zircons represent the only geological record we have for that period of earth history and thus the stories they tell take precedence over myths that arose in the absence of observational evidence."

It appears that there may have been two similar earths before and after the Late Heavy Bombardment, with a wild, crust-buckling ride in between. What happened in between is where JT and I walked. Nine in the morning, ten in the morning, then eleven, the black became uncomfortably hot as the sun passed to the top of the sky. The place didn't look at all ready for reseeding. Even if a fern spore blew in and somehow was getting a hold, within a week it would be covered again in a fresh new sheet of lava. We walked around hot

---

portions surviving through impacts spread over tens of millions of years. "Even under the most extreme conditions we imposed, earth would not have been completely sterilized by the bombardment," Abramov said when his findings came out in 2009. Meanwhile, geochemists at UCLA have found evidence that plate tectonics was present in the first 500 million years of earth's history. Tectonics implies a planet with mountains and basins believed to be necessary for life. Michelle Hopkins, lead author of the findings, stated, "In high school we are taught to see the [early] Earth as a red, hellish, molten-lava Earth. Now we're seeing a new picture, more like today, with continents, water, blue sky, blue ocean, much earlier than we thought."

spots where seeds blown in would have toasted and popped on ropes of lava galvanized gray, mineral skin glazed into dust.

"You ever been on lava like this?" I asked JT.

"No. Not this live."

"It's surprisingly hot," I said, reaching down to feel the iron-hot surface. My boots were starting to stink of melting rubber. "You want to duck into that kipuka over there?"

JT looked over to a small piece of forest in the distance. The lava had somehow spared about half an acre of *'ōhi'a* trees and feral papayas. It was surrounded on all sides by rock that had been on fire a few weeks ago. In the Hawaiian language, these are known as kipukas, places that survive.

"A little shade would be nice," he said, and we turned toward it.

Most Hawaiian lava flows, with their riverine fingers and bends, leave kipukas, saved by the fortune of viscosity and terrain. These kipukas become gene sources for eventual succession when seeds blow out of them onto cooled lava, niggling down in the cracks, starting over again. It's friendlier, swifter, and farther ahead on the evolutionary chain than reseeding via meteorite-carried microbes.

This kipuka was slightly sunken into the bottom of a gradual thousand-foot slope that had been a live lava cascade only weeks before. New rocks the size of flowerpots rolled out from under us as we slid down into the thicket. Through a burned-limb exterior, our boots crunched into dry leaves and then soil as we entered a small green glen. JT stripped his shirt and sank down against his pack, peering up into a light canopy of green leaves as he emptied a water bottle into his mouth.

"This apocalyptic enough for you?" he asked, his hat thrown on the ground.

I assumed his same position and let out a sigh as I leaned against my pack. "It's just right."

The Late Heavy Bombardment appears to have been caused by Saturn and Jupiter entering a gravitational resonance with one another, which threw their orbits out of whack. As the two planets rapidly migrated farther out in the solar system, the asteroid belt was disrupted, cluttering space with wayward asteroids. I asked Doug Robertson if anything like that could happen again. He gave me a reassuring smile and said, "Not at all. The asteroid belt is quite stable."

Although we appear to be safe from total asteroid annihilation, Robertson told me that to really wreck things, sometimes all you need is just one. With at least 4,700 within orbital range of earth, it's something to be aware of. He said, "This is the critical thing a lot of geologists miss: They think dinosaurs were killed while the surviving species were on a Sunday school picnic. But after Chicxulub, almost everything was dead on the entire planet. To a first approximation nothing was alive with only a few rare exceptions, individuals that sheltered from the heat by being underground or underwater. A few mammals looked up from their little holes, along with a couple of snakes, some frogs and turtles. Plants came back fairly quickly from roots or seeds. So within a century or two it would look fairly normal, except there weren't any big animals around. It took ten, fifteen million years to evolve the big animals again. But it was a fairly well-populated planet not long afterward. Recall that in Australia two or three rabbits populated the whole continent in a couple of decades. Very small numbers of survivors would have repopulated in the timescale of a century or two. You can't see that in the fossil record. What you see is that the big ones are gone."

Based on the initial air temperature following a major impact, which could have been up to 1,200 degrees Fahrenheit, Robertson figures that being under three or four inches of soil may have meant survival.

"Think about what species would be underground or underwater," he said.

"Burrowing animals," I said.

"Right. Mammals, maybe birds. Turtles, salamanders, insects. Do you see where I'm going? What is alive today is descended from what would have been underground at the time. Roots would still be viable, and seeds would have been in the soil. These things survived."

"That's all you need is ten centimeters? And you're good?" I asked.

"Ten centimeters, and then you crawl out into a devastated world where you have to survive."

Robertson's model of complete surface destruction is not entirely accepted by the scientific community. On the other side of the argument, fossilized hadrosaur bones have been found in the desert of northwest New Mexico, and they date to 700,000 years after the Chicxulub impact, or at least after a layer of iridium covered the globe, believed to have come from that event. The presence of large two-legged herbivores that much later suggests that some dinosaurs made it through and actually lived for a very long time. Defending his dinosaurs-dead-in-a-day theory, Robertson is quick to point out that the bones could have been rearranged by geologic processes, redeposited in younger layers so that they appear to have been deposited more recently. "No articulated dinosaur skeletons have been found after the impact," he said. "Since the connective ligaments do not survive, reworked fossils always consist of disarticulated bones. Articulated skeletons are the only reliable indication of unreworked deposits."

James Fassett, a retired U.S. Geological Survey geologist in New Mexico, was the discoverer of these lingering hadrosaur bones. He and his colleagues tested concentrations of rare earth metals both in the fossils and in the rock around them, showing that the bones were deposited and fossilized in that same layer and were not reworked at a later date. He also found forty-three skeletal elements from one individual, so the dinosaur probably dropped in that very spot, perhaps the last of its kind long after Robertson would have all dinosaurs burned to a crisp.

Personally, I side with Fassett's picture for reasons not entirely scientific. Robertson's incandescent sky from pole to pole, every fleshy creature burned out on the ground like tanks in a desert war, is just too painful to cuddle up with. I'd rather see Fassett's sanctuary of hadrosaurs living in some half-diminished Eden, a river that survived, a forest intact, whatever was needed to get them through another 700,000 years. This is more reminiscent of how things work on this planet, not a singular, uncontestable change, but more labyrinthine, kipukas scattered among the desolation.

In Fassett's defense, huge impacts do not always result in mass extinctions. Cataclysms may not be as instantly and naturally destructive as they appear. What is more likely than Robertson's totality model of sixty-five million years ago is a mosaic of die-offs, an asteroid impact here, flood-basalt volcanoes there, and whatever else happened to degrade the earth's environment step-by-step until finally dinosaurs no longer existed.

I woke after half an hour, sat up, and looked around. Dappled shade. Hot outside. The world was seared around us while trees held just a bit of cool. JT snored lightly, a camera rising and falling on his chest. I quietly pulled on boots.

Outside the kipuka, I walked through a calamitous glare. The island of trees was within a wide moat of high-silicate lava, wretched clinkers fallen all over each other like hills of mangled scrap iron. I climbed to the top of a pile. Shirt unbuttoned, skin fish-belly white, I stood out in the black expanse taking in the scope of lava. The rest of the forest that had existed here had been utterly eradicated. Volcanoes do far more than just cover things over. As Steve Sparks mentioned, eruptions tend to load the atmosphere with sulfuric droplets and other aerosols. Explosive, gaseous volcanoes like those going off in Iceland lately with twenty-mile-tall plumes laced in lightning tend to go straight to the stratosphere, a level of atmosphere that does not readily mix with the rest of the convective air masses we think of as weather. What volcanoes put up tends to stay circulating for several years. Like any swift chemical change to the layers above us, historically large eruptions have thrown global climates for a loop with a mix of both cold and hot reactions. The Eldgjá eruption in Iceland in A.D. 934 shot gas and lava out of a fissure fifty-seven kilometers long, putting up at least 185 megatons of sulfur dioxide. This eruption was followed by intensely hot summers in China and then intensely cold winters that led to crop failures that may have precipitated the collapse of the Chinese central government. A smaller eruption from the same region of Iceland in 1783 killed most of the country's livestock with a fog of fluorine and sulfur dioxide. As a result, over the next two years about ten thousand people there died from starvation. Comparatively, these Hawaiian flows are barely harmful, their output of sulfur dioxide (which reacts with sunlight, water, oxygen, and airborne dust to form suspended sulfuric acid) becomes more of a local breathing problem than a worldwide climate changer.

JT emerged from the shade buttoning his shirt. He scrambled up to his own haggard lava top with a camera on his back.

"Maybe we should set camp here," I said, gesturing at the shade behind us.

"In the kipuka?"

He was obviously feeling refreshed, ready to keep going.

"Come on, Wilderness," JT said. "Let's at least get up this slope."

The lava-fall slope to which he referred rose another thousand feet toward the center of the rift zone. We climbed through astonishing shapes, a single sculpture unbroken for miles, shapes of the newborn world, snail shells and the scrolls of violins. There had to be names for these things, I felt, the articulated ear of a bat and the tiny, hard metal tongue of an orchid. This was the beginning of the world, not the end. The end happened already, that earth passed beneath our feet, and now we walked on the shapes of rebirth—brutal, elemental, and as intricately shaped as life itself.

I held my breath across crags emitting sulfur steam, following JT as we moved through gossamer, throat-scratching air issuing from a long crack in the ground. The air wasn't so thick that we had to pull out masks we had brought along. Those were in case something was boiling up and closing our throats, as these thousand-degree SO_2 eruptions sometimes do. We came to the source of the smoke, finding a yellow mineralized gash big enough you could drop in a couple transit buses. The bottom was a rock-tumbled vent. Powdered silica blew around inside, swirling up little dust devils like eddies of glitter.

"What the hell is that thing?" I asked.

"One of Pele's many vents," JT said.

Out in clear air again, we heard a helicopter chopping over our heads at about five hundred feet. Did I mention the helicopters? By midday they were everywhere, like flies. Peering down through a Plexiglas bubble, tourists were paying a hundred bucks a pop to get firsthand views of the lava shield.

Look, honey, there are two guys out there. What in tarnation do you think they are doing?

Sir, we have come looking for the dangerously beating heart of the earth.

Could these people in the helicopters actually see us trudging up the slope? Probably not. There was too much land and smoke below them. I've been up in one of those lava-tour helicopters, and I can attest from that vantage the black and crenulated shield coming off this side of Mauna Loa looks like a limitless and impenetrable hell. When I first saw it, I wanted nothing more than to get out and walk across it to lose myself in the heartbreaking beauty of a world reborn.

Lava rock wound around long green kipukas where asphalt streets and house lots were briefly exposed, an island of a subdivision. Street signs stood nonsensically where cars no longer drive. We climbed into late afternoon until we stopped off the tip of a kipuka marked with stands of blackened dead 'ōhi'a trees. The last of the day's helicopters were buzzing around a new plume of smoke rising in the north.

"I bet there's something going on up there," he said.

"An outbreak," I said.

JT folded his arms watching helicopters circle a couple miles away. "You got plans for tonight?" he asked.

We finished dinner and just after sunset planted our big packs atop the highest coal-black knob as markers, so we might be able to find camp again in the dark. I grabbed my stick as we headed for where the helicopters had been buzzing. The sun was down, and aerial tourists had all gone home. We'd have the east rift to ourselves. We had pared down to day packs, and JT hauled a strap-on tripod as we hiked around smoking mouths of rock.

In dusk light I touched a cracked rim, rock hot enough I could only hold my fingertips lightly against it before pulling them back. Spit bubbled away, dried in seconds. "It's fresh around here."

"Last night's flow," he said.

It was like walking through shattered plate-glass windows, the remains of sudden decompressions and gas bubbles the size of Volkswagens. At the last skim of light, we began to see flickers in the distance. As we came closer, the land ahead glowed, lanterns raising.

JT was already moving faster toward it until I could barely see his form, a shadow against the red spectrum.

I switched on my headlamp and made the best straight line I could.

Soon, we did not need lights to see.

JT laughed with disbelief, his tripod unsaddled and planted on hot ground in one swift move.

"Look at this place," he said, his face lit up in the candy-apple glow of lava. The moving front must have been a hundred yards long and as contorted as a coastline with bays and jutting heads. No flame, it was half liquid pouring over yesterday's flow, picking up shadowy pieces of itself and carrying them along. I walked into a wall of heat, seeing shadows tear open and spill out ropy, bright entrails. A table of hardened lava reared back like a sinking rowboat, revealing a white-hot and dripping keel.

Moving at a few inches per second, this was not like water. It was luminous honey that hardened as it touched air. Liquid suns spilled and spread, covering ground that looked as if it had only been created the night before. This solid ground we stood on was riddled with finger-width fissures, and inside each was a deep solar glow. It had been liquid the night before.

I shifted from foot to foot, taking in the full circle of this dark, glow-cracked plain being overcome by slow, shin-high bellies of lava. The profusion of creativity struck me, so easily replaced as yesterday's intricate forms were swallowed by today's. Here in the rock was the shape of a crocodile, the first banyan tree, a shoulder, a hip. In the next moments, they were buried as bright rivers quickened, then bunched together in dark knots. Ten feet is as close as you could get comfortably. Five feet felt as if the skin on your face would start blistering. I darted in closer, four feet, three, close enough that the bald heat instantly stripped sweat from my face. I could barely keep my eyes open, seeing up close these brilliant folds, tiny bubbles touching the surface and stretching out. One second, maybe two. I spun and high-stepped into the cooler night. Having held my breath for fear of burning my lungs, I took in sea air blown up here from several miles away.

You could walk right along with the front of lava, backing up as you went. I moved laterally along its leading edge following protuberant lobes, globs, and globes, stepping in and out of their paths.

Glutinous surfaces slurred with a metallic hiss and they smelled of a blacksmith's shop without the oil. I reached out my papaya stick—its tip already blackened by someone else doing the same thing miles from here—and I poked it in. With the resistance of skin, the lava gave way, and I opened a small hole into a brilliant interior, almost too hot and bright to look into.

The heat inside of the earth, which drives both tectonics and volcanism, as well as our protective magnetic field, reaches temperatures nearing 12,000 degrees Fahrenheit. This heat exists because of earlier bolide impacts, mostly a by-product of the Late Heavy Bombardment, and other collisions from our primitive solar system as far back as five billion years ago. These impacts triggered nuclear reactions, and unstable elements formed deep within the earth where their half-lives are slowly burning out. Radioactive decay causes heat, and that is why the earth's interior is hot. (The three main sources of this heat are thorium with a 14.5-billion-year half-life, uranium at 4.5 billion years, and potassium at 1.3 billion years.)

Although many earth scientists have warned me away from connections between bolide impacts and volcanism, I see them as the same in spirit—incandescence. This heat of collisions is kept neatly inside the sunlit shell of our planet, magma breaking through only in those places that are thin or under extreme pressure. The pressure under Hawai'i in the middle of the Pacific Plate has long been a mystery, a lone hot spot where it looks as if a needle had poked a hole into a dome of superheated rock.* Whatever it may be, it is a window into the fire that drives this planet. When you punch a stick through its cooling, swollen surface, you open up a hole as bright as the sun. It may sound like a ridiculously silly thing to do, but put in the same situation you'd probably do the same. You'd want to interact with the inside of the planet if it was spilling out at your feet.

I poked a second hole, pulling out the stick as a flame streaked out behind it. I had turned on a blowtorch, gases hitting oxygen around 1,000 degrees Fahrenheit. The hole closed back in, swallowing itself as the wound healed and the sluggish, molten curve continued advancing.

JT let out a sudden yelp and was hopping away from his tripod on one foot as if he'd just found a live coal in his boot.

"What?"

Dancing through obscenities, he said, "It was glowing under me. I didn't see it."

He'd been crouched at his viewfinder flicking f-stops, not noticing his right boot on a hairline fracture blushing orange inside. The heat had come up through his boot all at once. He didn't seem injured as he walked it off. I kept going along this spreading lake that hissed, sputtered, and growled. A surge came up for about a minute with metal dragging against itself, squealing and creaking, yesterday's flow broken and pushed ahead like scrap iron raked across a car hood.

---

* Dumbbell-shaped heat lenses are seated beneath the crust on opposite sides of the planet, each covering millions of square miles. The Hawaiian Islands are centered on one, and the other, at the antipode, is beneath Africa. These heat wells have been alternatively proposed to be by-products of being insulated beneath two large tectonic plates, one being the Pacific, the other Africa. Or these could be a memory of some major strike, perhaps one imagined from the earliest earth, a Mars-sized collision thought to have formed the moon by knocking it out of the earth. Heat memory from such an event would still exist today, considering the half-life of thorium at 14.5 billion years.

Then it stilled and returned to glassy pops and whistles snapping out of a resinous, moving surface. I paused at a syrupy battalion of flows overcoming a rock hump. Breaking into fat fingers, lava rolled over itself and poured down the other side.

I fished a 1977 penny from my shirt pocket. I had brought it up with the idea of tossing it into the lava as if into a wishing well. It was hot in my hand, almost too hot to hold. When I threw it onto a fat pillow unfolding ten feet away, I thought it would land as if in pudding. Instead, it made a distinct ping and bounced right off. It's metal, I realized. It hardens like weld. This is what lies beneath your feet, not the soft rind of soil or two-thirds water we are so fond of. This is the earth.

If this kept flowing for a few days, the lava junkies would show up. And then tourists would start hearing. Pretty soon it would be a nightly pilgrimage. You would find offerings around here, a stone wrapped in a banana leaf, a bottle of gin, the neat coil of someone's necklace. Facing the elemental fuel of life itself, you have to offer something.

Lights cast in the dark. JT stood on a knoll in the stars panning around while I held my headlamp in an outstretched hand. The glow was behind us, out of sight, and now there was nothing but stars over an ink-black earth. Apparently, we'd both been wrong about where camp was.

"I don't remember this," I said as I looked down into a hole big enough to swallow a house. My light traced around its inside rim, illuminating curls of smoke ghosting upward.

"I think we're too far east," I said.

"East?" he said. "I was thinking south."

With the visibility range of a headlamp, nothing made much sense, black arms folded all over each other no matter which way we went. We walked in and out of patternless rock, occasionally encountering the last bones of trees where kipukas had burned out, but none were the kipuka we had used as a landmark for camp. The lava beneath us had cooled until it held no warmth at all, a flow months, if not years, old.

"We should stop and wait for morning; this is stupid," JT said.

"Yeah," I said. I was tired. My feet hurt.

"I vote we get a little sleep," JT added.

I dropped my day pack. "How about here?"

JT found a nook in the ground, a place to put his shoulders. I picked a back-sized convexity, a slight dome, and lay across it. Glass pricked through a light coat as I let my weight down, proffering myself like a sacrifice, my chest peaked toward the stars, arms across my chest to keep me warm. I closed my eyes.

An hour later, my eyes flickered open. Chilled by the cool tropical evening I looked around through a half dream. A lopsided egg of the moon was coming out of the sea miles away. I sat up and blinked at its battered face.

"JT," I said.

"Unh?"

"The moon."

"Ah," he said with a sigh as he sat up. "The moon."

There was enough light we could make out the lay of lava and distant kipukas. We were in a land of goblins. We got up and moved through them, and we found camp around sunrise. Unfurling our sleeping bags and dropping onto them, we slept out the morning while the sun rose and the grinning moon diminished overhead.

On open lava, biotic succession generally starts with crickets and spiders. And then you see small green flames of crack ferns, *Polypodium pellucidum*, darting up through the crunch. Moisture gathers, a little rainwater in hairline fractures where rhizomes break down glass and iron to pockets of primitive soil. You start to find black-berried creepers of the coffee family, *Coprosma ernodeoides*, and gnarled, red-flowered 'ōhi'a shrubs, *Metrosideros polymorpha*.

On the other side of the puffing hole of the Pu'u 'O'o crater, beyond the billowing, dike-fed vents of the east rift, and miles beyond the just-birthed lava, biotic succession climbed all over itself. Young, wet forests tangled up the sky, oversexed and colonizing lava flows a century or two old. You could only move by trail or road through this kind of jungle. On gravel tracks we entered what you might call a subdivision, but not in any suburban sense. It was mostly impenetrable walls of sun-grappling vegetation. Driveways had chains pulled across them, "No Trespassing" signs posted out front, and it looked as if they'd close in a matter of years if they weren't kept clear.

My gaze fell on a car junked along the side of the road. I nearly didn't see it, half devoured by plant life as it was. It looked like an old Japanese beater, make or model I couldn't tell underneath the foliage. It had broken down or was just left on the side of the road for too long, where it was becoming a topiary. Looking through its missing windshield and seeing its steering wheel and nematode-eaten backrest inside, I understood how hungry life is to get back into action, *Myoporum* flowers pink and dainty, *Cheirodendron* tangled through open windows.

We caught a ride and were dropped off several miles farther where the forest had grown much older on its bed of lava, soil now ten feet deep. This wasn't the mad scramble of subdivision succession. It was what you might call climax growth, a jungle having reached its native state. Before hopping the fence and entering its layers of trunks and vines, we took several minutes to pick seeds out of our socks. Using a wire brush, as we'd been told to do, we scrubbed our boot soles. We had signed a paper saying we'd adhere to strict regulations. This would prevent us from tracking exotic plants from the subdivision into native rain forest protected back off this road. It was one of the last large intact native tree-fern forests on the island. As a federal reserve, this detached unit of the National Park Service was usually visited only by scientists or workers sent out to shoot any feral pigs or cats that managed to get through the fence.* We had washed our bags

---

* Pigs are not native to the Hawaiian Islands. In these wet-forest ecosystems, they root out the bottoms of tree ferns, pushing them over with their heads to get at more edible roots or insects in the soil. Undermined tree ferns fall, leaving atypical pits in the floor

in town, and JT had shaken out his little yellow tent just to make sure we were clean.

This was a nine-thousand-acre refuge, a claim of native species on an island particularly hard-hit by recent extinctions. The woman who okayed our permit to go here, a sharp-eyed botanist sitting on the edge of her desk in soiled work clothes, had many questions.

"To my knowledge, nobody has ever requested permission to go in there. . . . I can send you to places you wouldn't know the difference where you can walk on a trail. . . . Tell me again why you need to be in this one forest."

"I photograph significant ecosystems around the world," JT said. "This is one."

"And you?"

"I write about the same kinds of places."

That was in lieu of a longer answer that might have prevented our getting a permit: *I want to see what the end of the world looks like long after the smoke has cleared.*

"Are you going to use its name?" she asked.

"Absolutely not."

"Don't use mine either."

We got a permit stamped and signed and were given a little twist of wire to attach it to our packs.

We hopped the fence and slithered through branches and fern stipes into what looked like a single green organism with no top or bottom, twilight at noon. Twenty feet in, the road was gone. A couple thousand feet and I'd lost sight of JT. I stopped and listened through a slow mist, hearing the shutter click on his camera. Moving several feet ahead, I came into view of him taking pictures of a moss-hooded stump of a fallen tree. His camera lens was half an inch away from the image, a thread of sporophyte, tiny bent-neck crosier growing from the moss. And beside it a snail the size of an *a* on this page crept over the atlas of a fallen leaf.

---

of the rain forest. Those pits fill with water. Mosquitoes lay eggs. Swamps form. Seeds blown or walked in from other ecosystems take hold in new environmental conditions. The forest changes.

"This is an island," JT said. "There's an entire ecosystem on this stump."

JT lowered his camera and asked, "Can you tell bearings?"

I stopped. We were both wearing full packs, carrying in our camp. Every direction looked more or less the same, layers of canopy closing out the view. It wasn't so thick that you needed a machete, not like the primary succession on century-old lava. But this was amply crowded. I could only tell distance through windows, happenstances of vegetation where I could see five or fifty straight feet. One step in any direction, a window would close, another would open, as if we were in a hall of mirrors.

"North," I said, pointing in some direction.

"I was going to say over there," JT said, about five degrees off from me. He opened a compass. His arm pointed another twenty degrees to his favor, both of us wrong. We reset an eastward course toward the heart of this refuge, tangling through veils and streamers of plant life, stepping in and out windows. Pearl-colored fruits hung like eggs around our knees, drops of water knocked off as we passed.

"It's the anti-lava," JT announced as he pulled off his pack, dropped out of view, crawled through a hole, and dragged his gear up the other side. He grinned back at me and said, "There's nothing not alive anywhere."

Records from this forest put precipitation at two hundred inches of rain a year, seventeen of those inches coming from fog drip alone. Pillars of half-rotted tree ferns leaned against palatial 'ōhi'a trunks, using them for support to get another ten feet, crowding the great 'ōhi'a until it was nearly invisible behind fronds so big you could roll yourself up in one like a pig in a blanket. This was a plant kingdom entirely. There were no hog trails, no tiny paths of soft-nosed moles. The only two original native mammals in the Hawaiian Islands are bats and seals. The largest historic herbivores were ducks and geese. Humans didn't even arrive until two thousand years ago, when Polynesian colonists showed up in outriggers. I had never been in a landscape without mammals. Even in heavy Central American jungles you find paths. Nothing here was made for or by our kind.

For any species it is a long journey to get here. Those able to cross great distances on sea-drift or in the wind, and are tenacious enough to grow on cooled lava, diverged into subspecies on this most isolated island chain on earth, twenty-four hundred miles from the nearest continent. Surveys have logged approximately 8,759 endemic species from fungi to snails to fish ("endemic" means they evolved in this place alone and can be found nowhere else on earth). If there were ever a global kipuka, this island chain is one.

The refuge looked like no jungle I had ever seen, truly primordial, dominated by nonvascular plants, fern canopies as big as beach umbrellas. It was as if we'd been flung back 300 million years to the Carboniferous, era of giant insects and oxygen-saturated green—only without, thank God, the three-foot-long scorpions. Dominant trees were *'ōhi'a,* koa, and *'ōlapa.* Each tree was a piece of biogeography in itself, shingled with fungi and ferns, higher branch-cracks and knee-holes crammed with bromeliads and liverworts that unfurled above us.

When you look closely at any one tree, you see it is two or three different species grown all over one another, climbing one another's ladders toward the canopy. The *'ōlapa* tree, Ginseng family, bears numerous trunks that twist outward at every deck of sunlight. The *'ōhi'a* is a witch's nest of branches, though if it has the light, it shoots up straight as a pencil fifteen feet around.

The dominant tree was not a tree at all but the tree fern, *Cibotium glaucum,* locally known as *hapu'u 'i'i.* Stalks stood as big around as you could reach, made of red hairy scruff and stipes longer than your body. I poked my hand into one at eye level, fingering back to its fibrous trunk, where it wasn't wood inside, but instead felt like cork and potting soil. From within, all manner of other plants were growing, the trunk of the tree fern acting as a nursery. Each big fern unfurled into a Corinthian head forming almost solid lime-green sub-canopies. This forest was rolling, living, climbing as if there were no tomorrow. These Hawaiian rain forests seem to have no climax, no genetic plateau as most ecosystems do. Plant surveys have been taken in scattered remnants, and sections of rot plugged up from the ground, and they show that the forest keeps rotating its restless array of species. Sometimes it is dominated by tree ferns, sometimes by 'ōhi'a trees or koa wood. Marshes open and close as if breathing. It is an evolution factory, a constant rumble among species that traveled far and worked hard to get here.

Tiny spores, seeds, and eggs were glued to the backs of leaves or dangled by silk under long, stiff fern struts. Our faces brushed through dainty webs of small spiders (unlike nearly palm-sized orb weavers we had seen in the sun gaps of the subdivision). There were no visible beetles, no sign of snakes, no croak of frogs. This was an odd place, a plant kingdom standing almost alone on a planet usually buzzing with insects, brush rustling with some bird or squirrel. There was not even ground to walk on, only a soup of shapes, ankle-sized holes, and fallen moss-backed timbers that made it seem as if we were traveling across sponge cake. We must have sounded like elephants, stymied, backtracking, trying other ways around huge fallen objects that nursed so much raw vegetation we couldn't even see what tree had originally been there. I referred every several minutes to my compass. Looking back, I could not find any sign that we had come through. Our every step closed in behind us, water pouring back into a hole.

If a major asteroid hit, or any high-speed cataclysm—humans included—you would hope there would be plenty of refuges like this to choose from to stock the future planet. Refuges ease the collapse, at least making the new start somewhat bearable. Fassett's hadrosaurs may have come from such a place, an island perhaps, a lost world saved from so many catastrophes. Our equivalent now is boreal forest in Siberia or Alaska, a succulent desert in Africa or Mexico. Prairie remnants, everglades, bottomless lakes, sand hills, tide pools, flyway islands, great lakes. You'd want as many of these reserves as possible. Maybe this is the unconscious drive for the botanists who protect this place, or the many others you meet tending to small sanctuaries such as this. We know how much can be lost in this world, and even more important, how much survivors count.

JT made a bed of dead fronds, leveling a space for his little yellow tent. I threw down under a knurl of tree roots that looked like a mossy troll house crammed between fat, shaggy roots. Before dropping my rain bivy and the rest of my gear into place, I peered into windows and pitch-dark places inside the tree's base. A Medusa-headed canopy rose above. What looked like a single tree was made of at least three distinct species climbing one another for daylight. If I could have answered the refuge botanist properly when she asked why we needed to see this particular forest, I would have told her I was looking for what the world is like after its surface was cleaned as if with a blowtorch, then given unfettered rein to rebirth. I wanted to follow the thread all the way through a destroyed, molten earth and out the other side. This is what cataclysmic extinction looks like when life flows back in.

What was different in this reborn world? Mammals didn't survive. The world was resurfaced and reformed, and we weren't one of the things to come back. Even insects barely made it. In this plant-dominated world, hardly a vertebrate remains. It is one of the possible futures, abundance returns, the earth lives on, and we are not there to see it.

The light was going fast. By three in the afternoon, evening dim had already threaded through the wet-jungle floor. By five, JT's tent was a soft yellow ovum, the last thing we could see as molasses shadows pooled around us. I felt as if a cocoon were gently sealing me in. Darkness climbed every tree and fern from the ground up, until outside the forest night finally fell, blackness doubling on itself. We cooked a meal, our pot boiling on the blue flame of a can stove, the only illumination.

Up high, wind purled the canopy roof. Where we were, the air didn't move, as still as the inside of a tomb. I turned on my headlamp and shone it up, illuminating detritus of seeds and spores slowly falling.

"We're in a snow globe," JT said.

Looking up through white-lit tongues of bromeliads and impenetrable tiers of higher canopies, we were seeing what cataclysm wrought. The force of the living is more cunning than any devastation, ready to explode onto whatever it touches. Under fat limbs and umbrellas of tree ferns, life was spilling down, planting its own future.

9

Seas Boil

Her body flaked into transparent salt,
and her swift legs rooted to the ground.

—Anna Akhmatova

ATACAMA DESERT, CHILE

AT THE BOTTOM of a dusty grotto on the edge of the dry Atacama in northern Chile, I opened my eyes. The sound of a fly woke me. In the half-light, I looked up through a cavernous ceiling made of grotesquely eroded minerals and hardened silt. It wasn't much of a space, a crack, a cave. It was shelter from the sun. Bluish light filtered through fissures in what was not rock but a hard substance stewed out of the bottom of what once was a lake, an expansive saltwater body that no longer exists.

With the planet dipping back and forth to the sun, climates wrapping around one another, there is bound to be a place cooked to a crisp for 150 million years, a picture of the final end. I scanned until I found the fly buzzing from one nick of salt to the next. You can walk for days out here and not see a single living thing, no blade of grass, no ant. I glanced over at JT, the same JT I had crossed the east rift with in Hawai'i. It had been four months since we were in the fern-tree forest together. Our paths crossed again in South America. We had gotten a ride to the edge of the desert, what is considered the driest non-polar region in the world, western rain shadow of the Andes, the air wrung of moisture from a cold Antarctic current flowing off the coast. A friend of JT's living in the one-story-adobe oasis town of San Pedro had driven us out in his rust-bottomed Subaru. When the road turned to drifted sand, he let us off with enough water to walk for several days. The plan was to cross open desert, going up and over a barren salt range, and hit the road to a lithium mine on the other side, where we'd hitch out. Our water should last until then.

JT was asleep in his usual afternoon position, camera on his chest slowly rising and falling. He looked as if he had been crawling around in a salt mine, clothes and face chalked. I thought of waking him and pointing out the fly, an actual living thing, but he seemed at peace. I rose quietly and crawled through a weirdly shaped blowhole that opened into sheer white sunlight.

It looked heavenly outside, pearly, a landscape made entirely of salt and its mineral allies. I had to squint, remaining just within the shade. A crackled plain swept the horizon, ending fifty miles away in a chain of snow-dusted volcanoes rising nineteen thousand feet. It is one of the oldest deserts in the world. In parts of the Atacama, no measurable rain has fallen for centuries. The dunes I had once walked in Mexico were comparatively lush, at least a few inches of rain every year or two between droughts, enough that vibrant verbena flowers spread across the sand most Februaries. Here was nothing. You quickly begin to forget that you were ever alive.

At about eight thousand feet in elevation, we were on the edge of a salt pan, a shimmering expanse known as a salar. From space, these salars look like snow, pools of white strewn along the high Altiplano and into this broad, basin-riddled desert descending 160 miles to the coast. This was the textbook definition of uninhabitable. As a measure of this locale's extremity, probes sent to Mars looking for life were tested here in the Atacama. Little shovelfuls of soil were cooked to see if they showed any sign of microbial metabolism. They came up as empty here as they did on Mars, evidence, it would seem, of a lifeless planet.

In the fall of 2011, Jim Hansen, the NASA climate guru, announced, "If we stay on with business as usual, the southern U.S. will become almost uninhabitable." But what does it mean for a place to be "uninhabitable"?

In 2010, climate researchers from Australia's University of New South Wales and Purdue University in the United States released findings saying that by the year 2300 half the planet could be unsuitable for living by Hansen's same definition if greenhouse gas continued being produced at current levels. The study took into account longer time periods than most climate models address, three hundred years being a relative stretch of the imagination. And rather than focusing on the usual political and economic effects of climate change, researchers looked at how much the human body could actually withstand. Like most mammals, humans would experience potentially lethal levels of heat stress with an 11- to 12-degree Celsius rise in temperature by 2300. In this scenario, large portions of the earth would simply be off-limits to our kind without substantial adaptations. There is some question whether this scenario is even possible, as additional heat at that extreme would increase cumulus convection, and with it more cloud cover, more shade, potentially more precipitation. But lifeless it would not be. Again, back to Edward O. Wilson, who said we could take the world back down to the microbial level if we wished to explore such horrors.

I turned back inside and swept out a place for my shoulders. It was like trying to rest on a floor covered with toys, pieces of friable crust of semi-rock fallen from above. I went down, shuffled shoulders flat, and closed my eyes. In this rarefied silence, a sound so hollow it seemed audible, the fly began buzzing again. My eyes opened. It was turning circles above us, probably excited by our presence, stirred by the smell of our salt-greased bodies.

Without opening his eyes, JT said, "The fly."

"Yeah," I said.

"We are not alone."

Leaving our shelter was like climbing out of a hole in the face of the sun. Our shadows moved crisp across a rock-hard surface, salt cemented into bizarre formations. Down below in the salar the salt lay low and crinkled, as level as a billiard table where every step sounded like you were walking across dried armadillo skins. Up here, we walked on hard, broken halite. We were on an uplift, a tectonic surge that brought up not rock or soil or even much sand but crusty cocktails of sulfates and salts, glauberite forming flowerlike crystals mixed with potassium, calcium, and magnesium that long ago leached from shallow brines. We booted our way up evaporated minerals dropped about thirty million years ago. Even at this moment the cycle was still turning, snowmelt runoff from the high ranges pooling into evaporating lakes, reservoirs turning to ponds that haze back into the air, leaving so much evaporated matter mixed with volcanic mud you could make mountains out of it. As I climbed, I carried water strapped tight to my back, thinking if this were sheer deadness, it would have been exhausting. But it was a different kind of life, a mineral world, a form of endurance beyond our fragile skins.

Even when biology is removed, every fossil hidden, you would see the mark of life on this planet in its minerals. When a planet is born from interstellar dust it has about twelve refractory minerals, those resistant to decomposition by heat, pressure, or chemical attack. By the time it is complete with asteroid accretion and finally volcanic activity, about 1,500 different minerals are present. The earth has at least 4,300 species of mineral. This high number is unique in the solar system, a function of biological processes such as photosynthesis that releases oxygen which chemically bonds with almost every element, creating new minerals.

Most of what I could see as I walked through the Atacama was one of the older mineral species, halite, a form of sodium chloride, rock salt, NaCl. It is known from the formation of the solar system. A meteorite that landed on earth was dated to 4.57 billion years old and inside it was found original halite that must have formed within two million years of the start of our solar system. Knowing the age of this mineral made it feel as if I were walking in the company of extreme time. The dunes I had once walked were made of sand dating back millions of years, the broken-down remains of a recent planet swept into an ephemeral desert. The Atacama, on the other hand, felt like the oldest memories of earth.

In my cheek I milked a wad of coca leaves, JT's recommendation.

He'd picked up a fresh bag in San Pedro at the market. He swore you could go forever on this stuff, having spent time on the Altiplano and down south in Patagonia crossing large pieces of land by foot and horseback with a hunk of coca in his mouth. I wanted to seem like a good *compañero* and was happy for the faint, tingling buzz it offered, but it wasn't the best-tasting thing I'd ever encountered, like a cheek full of lawn leaves.

As if it were liquid, salt oozed from the ground in crusts and pools and shattered bubbles. But it was not liquid, not in the way we usually think. It flowed, but so slowly you'd have to wait lifetimes to notice the change. I cracked off a piece and nipped it with my teeth, rich taste, minerals, bitter hint of bicarbonate. It went well with the greenish juice I kept swallowing. Maybe I could see why JT liked coca. It gave me a point of focus as I stopped to catch my breath and look across the breadth of desolation below, no particular place for the eye to land, only a distant train of volcanoes rimming the Argentine border. I'd found the fundamental earth, the essence of this planet that bore us, only now I was down to the grain, the mineral. Still, it was a patterned landscape, old shorelines and layers of deposit laid into the ground from great lakes forming and receding over millions of years. Life was defined not by green breathing organisms but by larger forces continuing to turn around each other, cycles and rhythms. It was a different kind of vitality. The last vitality on earth.

We weren't carrying a lot of weight, a fair seventy pounds, but I'd been eating bonbons with the kids at home, and JT had been in St. Thomas working on his sailing. Neither of us was in the best shape. When I climbed a short slope, it took a moment to get my breath back. Winding up here together had been fortuitous. Leaving Hawai'i four months ago and seeing we were going to be in South America at the same time, we cooked up this journey. JT had been down south running the Rio Baker in Patagonia, trying to stop those same dams, while I was looking for a planet burned down to its sere end, cooked naked by the sun. However far out, there is a certain inevitability to this end. The sun is gradually becoming brighter. A billion years from now, it will have brightened by 11 percent, potentially raising temperatures by 120 degrees Fahrenheit.*

From the top of this spine we could get a witheringly good look around, pale depression of Llano de la Paciencia on one side, the shimmering white waste of the Salar de Atacama on the other. Somewhere ahead was the lithium mine with a road we were heading for a few days away.

There wasn't really a summit at the top. It was a row of halite steeples the shape of miniature Matterhorns thumbed into the sky. You get it in your head you're moving, stopping only to record oddities along the way, a note about a hoodooed shape, a photograph of glass salt spilling from a crack. You become a rover, a probe dropped onto another planet. I went down on my knees, licked a clear and eroded hunk of ground. How often would you get the chance, really? I bit off little tips, beginning to distinguish mineral properties, some places nipped off clean as table salt, others, like eating dry foam, the slightly sweet taste of arsenic.

"Listen to this," JT said, flicking a white bell of a salt bubble with his fingernail. It sounded like the high end of a xylophone.

"It's harder than I thought it'd be," I said.

"Hard as concrete," JT said. "You don't want to wipe your ass with this."

This could have been the floor of the Pacific boiled away long ago. Even the corpses of whales and stranded ships were no longer around, the world soundly ended. Snow-tipped volcanoes soaring off the horizon could have been the Hawaiian Islands standing over their baked ocean floor. We'd make a bright planet in the end, snowflake-colored basins covering more than 70 percent of the globe where seas once were. Only it wouldn't look exactly like this; there'd be a lot more gypsum squeezed out of the ground, the result of different mineral content in the oceans than what you find in these evapo-

* While the sun has continually brightened, steadily increasing by 30 percent during earth's history, temperatures have remained within a habitable range irrespective of that change. This is due to chemical balances in the atmosphere and oceans, partly maintained by the balancing respiration of living organisms.

rated inland lakes of South America's inside hip. The skies would be significantly different, too, more like Venus. All that water would have to go somewhere, simmered into the atmosphere, filling it with vapor, creating runaway greenhouse conditions at a cloud-boiling 900 degrees Fahrenheit (estimated temperature if earth's oceans became atmospheric). Heading to the opposite extreme, a planet like ours could become more like Mars, the atmosphere stripped away, liquid water either whisked off the planet or frozen into the ground as the inside of the planet cools and the magnetosphere dwindles until almost gone.

The Atacama has been generally lifeless for much longer than humans have existed, skies thin and roasting by day, letting out heat at night giving way to an uninsulated cold. At the end of the day I found JT stopped on a pie-crust dome of evaporites pierced by cave-like holes leading straight into the ground. Color was fading to gray.

"Camp?" JT said.

"Camp," I said.

We dropped packs. All around us were what looked like pouting blowholes and sunken, in-turned navels. Some were too small to put your hand inside, others opened into body-sized cavities; we couldn't see the bottom of any of them.

"Womb-like desert feature," JT said as he drew out his camera.

This is how he and I traveled, sniffing the ground with our pens and cameras, seeing what makes one spot distinct from another, milky bubbles mixed into countless horned helmets of salt, ancient battlements and bishops' hats. You wouldn't want to rush it, no need dropping in on an alien world if you can't stop and poke around.

I was glad to have some flat spots so we could put up our tents. This was soft ground compared with the stretches we'd been climbing earlier, where it was difficult just to find a place to sit. This was more like chalk.

For most of the day we'd been hearing crinkles and cracks from salt expanding in the sun. Now it was quiet and stars swept the sky, air touched with an early winter chill. Like Mauna Kea, the Atacama is dotted with telescope installations, one of the clearest skies you'll ever find on earth, though none were anywhere near us. The view from here is toward the bright center of our spiral galaxy, rather than out toward the last fading arms we see from the Northern Hemisphere. From this side, we were looking into a region of space where a powerful black hole is believed to be sucking in stars and gas clouds, piling them up around the swirling center of the Milky Way. It's as good a view of our edge-on galaxy you'll find anywhere on earth. There were dark spaces overhead, apertures enwreathed by stellar gas clouds, their insides starless. One was so black I thought it the blackest place I had ever seen in a naked night sky. It looked like a hole you could climb through, an exit from this world, entrance into another.

Dinner was salami, cheese, and fresh garlic bitten off and chewed. Meal done, coats zipped up, hats pulled down, we stuffed hands in our pockets and watched the sky turn. Quiet voices, wandering conversation, we were two men on a picnic, reclining in the grass under the shade of an apple tree, looking up and counting clouds. Only there was no grass and no tree, and the clouds we counted were light-years away.

"I'd like to navigate by stars across open water," JT said. "No compass or map. What's it from Fiji to Hawai'i, two thousand miles? You could do that. Get yourself an outrigger."

"Me?" I said.

"No, me. But you'd pick it up pretty fast."

Since we'd last been out together, JT had gotten his captain's license for open-ocean sailing, a lifelong goal of his. As we watched stars slowly shift, JT imagined he was navigating across the Pacific the way ancient Polynesian mariners did when it is said you could tell a navigator on the boat by his tired eyes, nothing but staring at stars all night long. As he thought of those nights, all others asleep and nothing but the docile sound of water while he remained on watch, he remembered the feel of a boat underneath him, the sound of water.

Rhapsodizing about how you catch the wind on open water, he

reached up his hands and said, "You turn the vessel's head into the wind so as to bring up the wind on the opposite side, and it turns you."

As he said this, I saw the hull of his boat passing through water overhead. In my mind's eye we were lying on the bottom of a lake thirty million years ago watching his vessel carve a turn above us, its wake rippling the stars.

We woke on a different planet, positioning ourselves for the first event of the day. The sun exploded through the peaked skyline of resting volcanoes in the east. Sitting among miniaturized crone-headed peaks of salt formations, JT sipped his maté and squinted into the light. He offered a pull. I took his cup and touched the metal straw as if to stir.

"Don't touch my fuckin' *bombilla*," JT snapped.

I hadn't been back in South America long enough to remember that you don't stir someone else's maté. I never liked the drink, anyway. Taking a polite sip, I thought it tasted like a vacant lot cut up and boiled.

"Thank you," I said, handing it back.

"You should drink more maté, it's good for you."

"I don't like coca leaves either," I said.

Later as we packed up, JT reached into his pouch and pulled out a pinch of coca as brittle as bay leaves. He looked at me. I said yes. I crammed brittles between my cheek and gum, chewing them into place. He then pulled out a charcoal-gray ball hard as clay and made of fire ash and soda.

"Devil's nut," he said. "It'll help."

I scraped a bite off the ball, wincing at its bitterness, my salivary glands leaping to attention. I wiped spit from the corner of my mouth.

JT smiled and said, "You'll like it."

The bicorbonate softens the astringent flavor a little and activates the alkaloids in the leaf, enhancing its numbing effect, I noticed.

A chalky landscape grew around us, some places white as egg-shell. Dawn had been pink and so had the salt, but now colors washed away as if cleaned and hung on the line. The sun was like a wind pushing against us. We tried walking south, but the sun pushed us west. We kept its light at our backs, shielding our faces. As light curved through the morning sky, we turned with it, entering the interior of this narrow range where we moved around sinkholes, their mouths revealing razor-rippled walls dropping twenty or thirty feet to inexplicably flat bottoms. Salt wells, saline deposits. We passed around their crumbling edges until there was more hole than land, some with floors unseen, bottomless, it seemed.

JT lifted a piece of crust the size of a bread loaf.

"Three . . . two . . . one . . . ," he said, and he threw it into one of these cavities.

One . . .

Two . . .

Three . . .

A sound popped back at us, a low thud like a heavy vase dropped on concrete.

"You wouldn't want to fall into that," JT concluded.

Standing across the hole from me, JT looked downright Bedouin even in a red ball cap. One thing that man does well is fit into his landscape. He'd bought a light black scarf in San Pedro, and it was now his everything, wrapped around his head and shoulders as if he were crossing the Sahara. It did a good job keeping the sun off. This is why I travel with smart people. They know how to piece it all together from what they can find, and somehow it works.

JT went ahead and crawled down the steeply sloped edge belly down, sticking his camera out as far as he could reach over one of these sinkholes, something I frankly would not have thought of doing. His shutter clicked down deep. He shimmied back up and flicked on the screen.

"Holy crap," I said, because I had no idea what I would see. It was a perfect circle, a plug sixty feet deep daggered with salt all the way up the walls, some kind of dark lord's castle, only inside out. You'd be shredded before you hit bottom.

"Talk about salt in the wounds," he said. "You definitely don't want to fall in one of these."

So we didn't. It was easy enough to walk bony high ground the texture of rough bricks and kernels of pale minerals you could stomp on and crush. We stretched out along bridges, dipping through swales. There was a bush. A dead bush. I didn't know dead for how long, years maybe, a genetic refugee from one of those odd rains. It looked petrified. Its leaves had withered back into tiny fists, no species I had ever known.

The Atacama does not feel like one of those deserts that would turn on a dime if rains began falling. A bush might spring up on this salt range, but only one for miles in any direction. It's not like the Sahara with a hyperarid core that would return to grasslands and marshes with a reintroduction of monsoons. This place has been mostly hyperarid for 150 million years. It is permanently scorched earth, a cigarette burn on the skin of South America. I'd seen one fly, one guanaco leg mummified and mineralized with some hide still on it, and this dead wire of this bush. As far as life, this was it.

I was glad for it not to be summer and thirty degrees warmer, but this low sun blistered through everything. I was used to having some trees at this elevation, some shade at least. The only shade was down in these sinkholes. We found one we could climb into. Leaving packs at the top, we climbed carefully, no undue risk taken, just our water with us, my notebook, and JT's shutter. Bulbous cauliflower formations hung from walls, grown into brain-like pillows under pieces of fallen debris. Salt life. The salt was marbled and tasted good. There we sat on butt-sized balconies at different heights, voices reverberating if we talked. A crescent of sunlight moved down on us and come noon our shade pinched out.

We climbed back out into the light, picked up packs, and traveled into a turret-riddled skyline. Sinkholes became smaller, then closed as we entered a region of pillars and high pedestals of ice-clear salt popping and creaking in the sun. Early afternoon, JT waved me over from a distance, far enough I had to consider the effort of stepping over mounds and gullies of saline shrapnel to get to him. He had dropped his pack in front of one of these crazed smokestacks and was sweeping around it with his camera as if it were a nude. I turned course from wherever I was heading. By the time I walked up to him, JT had set his camera aside and was seated, head in his lap. He had pulled his scarf over his body, making himself into a beetle, back curved against the light.

An arm appeared from under his beetle carapace, and he pointed at the salt formation. "Go put your ear up against it," he said.

I did as he said, stepping up a skirt of crystalline rubble where I put my ear to the surface. Inside it sounded like flowers made of glass slowly blooming. The salt outcrop was an acoustic antenna. I could hear sharp frazzles and snaps as sunlight entered the formation. Down deeper were twangs and pings within the thick crust. Our planet is a concentrator of minerals, particularly salt. When seawater seeps down through the crust and enters super-heated and pressurized layers, it leaves its salt behind, planting massive halite formations deep within the earth. Previously, concentrations of salt were believed to have come from the breakdown of rock within the earth's crust, but basic rocks simply do not have enough chlorine to account for the huge deposits of halite you might say this planet is famous for. According to the hydrothermal salt theory, you'd find salt formations on the surface in the company of volcanoes where underground water left its salt behind under heat and pressure. This turns out to be the case in the Atacama, as well as the Danakil Depression of Ethiopia, the Eritrean salts of the Afar Triangle in Djibouti, and the gypsum sands of White Sands National Monument in New Mexico. These salts are not just here from evaporating water bodies, they are likely part of circulations from within the planet.

JT uncurled from his beetle shelter.

"You heard it, too," he said.

We traveled on, sun bending our passage through the day as we walked into our own shadows, back on a south course by late afternoon, inching east. The shadow of our range stretched across the salar below. We'd gotten a little off during the day, our track more arced than we had planned, which put us in a landscape of steep slopes and hard toadstools of salt. It was a mess of a place. While JT was snapping pictures, I dropped my pen by accident, listened to it clatter down through coral-sharp, hardened, petrified muds. Well, damn. I had to go and get it, straddling a spine as if it were a horse, then crawling down and reaching through holes in a gully until I found the thing, lifting from a fissure and slipping it back in my pocket.

It was late in the day, and we were tired from the sun, glad to be getting down into the precipitous shade of our range. We jumped from one hardened toadstool to the next, each made of salt sediments cemented together, each taking our weight, thank goodness, because a fall would have been a thirty-foot pitch into another hundred-foot drop, someplace you'd be lucky to still be counting broken bones when you landed. Backtracking wasn't a sound alternative this late in the day, and after a while, you don't give a shit, you're just moving. We were trying to position ourselves to drop down to the salar on the other side of this range where in the morning we would make another open desert crossing, eventually out to the mine road. I was already starting to notice the weight of less water on our backs, now that we were two days, if we kept a healthy pace, from the nearest road, where we'd hopefully thumb a ride back to San Pedro, the oasis town with outdoor bars and bonfires at night, a place you start yearning for when you're walking in the desert.

Camp we took in back of a knolled ridge. By dawn we were up and gone, heading down toward the buckled plain of a salar, hurrying to reach the bottomland before sunrise. We didn't want to be caught with east-facing glare in our eyes while trying to climb, so we passed down packs, climbed down behind them through cathedrals of salt, the severe and glassy walls guiding us layer by layer until we hit the flat and expansive floor of the salar.

Ten miles wide at its thinnest and sixty miles long on its axis, we were crossing the salar on a diagonal. This was the remnant of a lake, or many lakes over and over for as long as this desert's been here and as long as climate fluctuations let water in every few tens of thousands of years. From within the salar, you don't see the larger ellipse of its shape, only the circle of what your eyes allow. As we moved across its crackled horizon, morning mirages flowed toward us. They began as distant ponds, little mirrors miles away. Heat of sunlight lifted beneath a layer of cool air from the night, the combination bending light into waves. Ponds of light became lakes, and we could see their edges spreading, joining. Lakes became a shallow sea perfectly reflecting mountains in the distance. We could see it moving, spreading. I would have sworn at that moment it was real, that if I had thrown a pebble, it would have rippled. We mounted a small hillock, a mere three feet in elevation, and on it we stood as if on a small island stranded in a silver ocean.

In the Atacama, I saw the future, when the sun eats up the last of its hydrogen and burns into its red-giant phase, big enough to cook life and clouds and oceans off this naked orb. It wouldn't be a fast process, not by our standards. Millions of years in the execution, our sky would finally be half filled by a sun the color of a red-hot moonrise. After that, the sun would probably collapse into a white dwarf, meanwhile blasting away its outer shells of gas into an explosive planetary nebula. I imagine that all of our minerals will pay off as we make a rainbow streak flaring off into space. We will be beautiful.

Even today's bright, thumb-sized, main-sequence sun—middle-aged and burning nicely—seemed a little much. In the midst of its crisp light, I swear I could hear or perhaps feel on my shoulders a torrential rain of neutrinos. Our route took us into zinc-white junk heaps of broken salt plates, complete chaos, sun beaming up inside our nostrils. It wasn't good walking. No trail, no clear path. The ground seemed bombed-out.

"It's surprisingly similar to walking through lava," JT said, more of a complaint than a comment.

I said, "Yeah, and I thought it was going to be more of a salt flat, you know . . . flat."

It was like walking through a coral reef. Ages of water flows and snowmelt floods from the volcanoes had sculpted the salar. When it dried, it contracted and buckled into table-sized pieces of hard ground hitched up all around us. The hardness was a function of salt content mixed with lake-bottom sediment forming a concrete as weirdly shaped as stag antlers. Our aim, for the sake of navigation across the salar, was a notch in a mountain, a strong V on the horizon about thirty miles across this jumped-up mess of salt.

The only thing out of place was a dead cow. We spotted it probably an eighth of a mile out, though for the preceding few minutes it had looked like some kind of fallen ghost, something we weren't sure was there or not.

"It's definitely something dead," JT said with genuine curiosity.

We had plenty of time to work ourselves up until we reached the pitiful thing, hooves splintered in the sun. You could kind of see a stain that had briefly been its body. JT said to it, "If you just get your wits together, you'll make it." Its hide was as hard as bone. I crouched and probed its eye sockets, tapped its drum skin with my pen. It was practically fossilized, skin and all.

JT took a few pictures, finally crawling right up to its skull for a macro view. Then he pushed his camera aside and without a word put his head on his hands and closed his eyes. At the head of this dead and solidified thing, he slept, as if the two of them were meant to be together, him and the cow, signs of life seeking signs of life.

In ten minutes, JT's nap was over, and I was still standing there, a statue over this cow, staring at the expanse around us. JT woke facedown, and groaned. He lifted to his knees, peered around as if for his bearings, and stood, where he started closing up his pack.

"I'm walking," he said.

And he did, his figure heading off toward the V in the mountain far away. Fifty yards out, the mirage began to take him, flickering around his heels. Thinking I should probably follow soon, though knowing on this salar I'd have plenty of time, I pulled a water bottle and unscrewed the lid. It was the only sound. Just about to take a drink, I again noticed the dead cow at my feet. Its skull had been untouched, left upside down with upper teeth shattered in their sockets. It seemed rude to drink in front of it. This animal had perished in what must have been a most unpleasant way. I reached out my bottle over the poor thing and poured some water. Perhaps I shouldn't have done it, wasteful, maybe fruitless. But I had to. These sorts of things matter. A quick stream spattered inside its skull. I didn't have to say anything after that, no *Anno Domini* or *I know it's a little late*. It was just a gesture of recognition, thinking this was no way to die, not even allowed to rot, just baked right into the ground, hollowed to a rind. I slugged back the rest of my water and shoved the empty bottle into its place in my gear. I looked ahead to see JT, who appeared to be twelve feet tall, his body eerily processed by mirages as he walked.

Noon was rather high and unavoidable. Mirages had all but fled. As I moved across the salar, I looked up for JT, seeing he'd mostly disappeared in the distance, his coca chugging him along at a good pace.

I lost track of time, not remembering what just happened other than the cadence of my steps. I stopped and looked behind me, seeing the faded rise of the salt range we had climbed and far behind that the skyline of the Cordillera Domeyko, a spur of the Andes on the other side of the patient valley. I looked ahead and JT was gone and the V was still in the mountain. I hardly cast enough of a shadow to reveal north, south, east, or west. There was no protection, a sensation that I treasure, especially when I feel plenty safe, some water on my back and a bead on my course.

It wasn't so much the heat as just the sun itself. The air was probably a comfortable 87 degrees Fahrenheit, but the sun was burning through me. I could feel the reason Southern Hemisphere skin cancer rates are higher than in the North. The ozone is measurably thinner on this side, a gauze rather than a shield. This elevation, now around seven thousand feet, magnified the sun through rarefied air, while the earth below bounced it back, a near 100 percent albedo. The salar is as good as an ice cap at putting radiation back into space, so you're walking through a column of that very radiation coming down and going back up, a straight line off this planet. It felt as if the sun were digesting me.

I looked down at my shoes. The ground was eating me, too, my soles rotted by salt and nitrates grinding step after step. I picked up the pace, chewing a little harder on my coca wad, juicing it up, though it felt drained after a couple hours, its tingle wearing off.

I caught up with JT. He had stopped and was sitting on his pack with his scarf pulled over his head. He was drawing a triangle on our map with a straightedge and a carpenter's pencil. The map was a ridiculously low-scale tourist brochure, the only one we had with us. I opened my mouth and coughed out my coca wad. It landed on the ground at my feet.

"More please," I said.

JT pinched me up some leaves.

"Devil's nut?"

"Mm-hmm," I said.

He rolled the noxious ball out of a dirty plastic bag, and I bit from it.

"*Gack*," I cried, holding in my spit.

"That'll wake you up."

In a dry wash aiming for the middle of the plain grew a bizarre-looking succulent like a heap of chubby green caterpillars, nothing like any other plant I had ever seen. Fly, guanaco leg, dead bush, green plant from Mars. It was a salt plant, actually, saline tolerant, living off moisture somewhere beneath the surface. Moisture? Here? I took it for what it was worth; the plant had enough of a shadow that I lay down and put my head under it. JT erected his yellow wing-shaped tent on something that resembled a flat spot and crawled in.

I swear that within a minute JT was cooing with a light snore.

That man never ceased to amaze me, especially as I stared at the sphere of the sky, thinking I should at least close my eyes. Not a hint of moisture trailed through the sky. It was a single color top to bottom, side to side, only washed out around the edges, where the last mirages hovered.

In twenty minutes JT was up. I was at least rested, good to get the noggin out of the sun for a little while.

Wake, walk, rest, drop. This is how you move across emptiness. Drink, note, stop, stare. Photograph, scribble, look at map, decide map is useless. Look at map again. No ants, no scavengers, no birds, no morsels of food. Only it wasn't emptiness. I was walking through rubble made of white-frosted plumes. Every step was a reach or a jump, pants ripping as they snagged on hard barbs of salt. The deeper into the salar we traveled, the more jumbled the ground became. At sunset, we were up and down through great blisters and car-hood-sized plates of tilted evaporite. None of this salt tasted like mountain salt. It was fine grained with heavy metals mixed in with salts grown into a carpet of mangy, knee-high crystals all crashed into one another through processes of expansion and contraction. It tasted like dirt and baking soda. The walking became slow, as in miles per day rather than per hour. We stepped up and over, weaving around, not one footfall like the last. I spit out my wad of coca and, as I walked I plucked a small orange from my pack, peeled it, and stashed the peel in my pocket. I bit into the taste of cool water and life, genetic memory entering me from far away, a different landscape washing bitter saliva from my mouth.

There was no place to sleep. That was pretty obvious. As the sun set and a cool evening flooded this low salar, we began to wonder how exactly we were going to drop a camp. You couldn't even sit, glass-sharp edges everywhere. It had been at least a mile since the last even remotely flat place. Now it would take a pickax to clear a space the size of your shoulders, or just a hole to kind of curl up and rest in. I had really thought this salar was going to be a pool table ten miles across, a mysterious land of mirrored earth and sky across which one would stumble in a coca-induced delirium. That's what I had hoped for, at least. Instead, this was work—up, down, over, losing daylight, crashing through foot holes. Headlamps came out, and we continued.

Note to future travelers of boiled-away seas: At least be ready to walk all night.

We found a low spot big enough for two. The Big Dipper was tapped out on the horizon and was swinging low, maybe midnight. Many miles south by southeast was the glow of the lithium mine. Cold had moved in, freezing now, the air absolutely still. We cooked something up for dinner, bundled warm, and as I slid into my bag for the night under stars, I smelled what I thought was the ocean.

Most of our water froze during the night, and when we left at dawn, still heading for the V in the mountain, we shrugged off the cold, shaking out sore limbs. A cat-scratch moon washed out in the sky as the sun rose just behind it. White light sheeted the desert, a scentless barrage of radiation that if nothing else got the blood pumping. Cold fingers stiff from stuffing gear warmed up right away.

That's when we stopped, finding a hole in the ground the size of a coffee can and in it, water. I dropped to a knee, scooped it up to my lips. It might have made you vomit. It tasted like ocean, only much stronger. Salt crystals had formed inside the small pool, delicate chains hanging in the water. This was where high snowmelt ends. It had flowed down from the volcanoes, sunk underground, and eventually settled beneath the salar, the lowest point on the landscape. This was the water that sustained my shade plant from the day before. Had it been beneath us the whole time? Had we taken a pickax to the ground making a place to sleep, would a fountain have welled up? I looked down into this one, a tiny cave in the ground decorated with suspended crystals. You couldn't drink it, I knew that. You'd live longer on seawater if you had to. And who knew what these heavy metals might do swirling in your liver?

We knew there was a chance of hitting groundwater somewhere in the middle of the salar, but the day before, while we studied our simple map, it seemed that finding a water body on this salar would be like trying to find a dime on a basketball court.

We'd found the dime.

That's when we looked up and saw the pink flamingos. Flying in a crucifix formation, and as rosy as any neon sunset, seven of them gabbled at one another as they passed overhead.

As if that weren't enough, I noticed morning mirages had started up, their impeachable silver tide rolling in.

"Flamingos," JT said, dumbfounded.

We knew this was a possibility.

We were in the range of the Andean flamingo, *Phoenicopterus andinus*. With curiously dipped bills, they strained microorganisms and goggle-eyed brine shrimps out of saltwater bodies that collect in these salars west of the Altiplano. Their presence wasn't out of the question, but it certainly seemed it should have been. In all my life I had never seen flamingos in flight, and they looked awkward and strange, like many-jointed bombers droning overhead. They landed in waves of pink, starting up morning chatter. We turned and walked on into the glare of the sun, layers stripped off, sleeves rolled up, while ice sloshed in our packs and mirages coalesced, enough sensory input to make a person go suddenly mad.

The ground became fragile. We tiptoed through antlered masses and frosted, glassy chrysalises of salt. We had entered a basin of primordial, organic shapes. These were natural patterns, crystalline repetitions that seem almost like an early, nonbreathing form of life. Never had I walked through such penetrating beauty. Snow-white billows looked like clouds wormed with delicate salt tubes. Fine rimes of evaporite filled with sunlight, their walls as friable as pastry shells, shy and hooded creatures, fringe of oyster, vent of octopus. From ahead, we heard the murmur of flamingoes. We came upon a braided, flowing stream wide as any river and inches deep. Flamingos stood among one another as if at cocktail hour. They had come

320

for morning feeding and the other mysterious routines kept by birds. Here they would graze as soon as the grain-sized shrimps started to move in the warmth of daylight.

I had not expected to see a living thing, and here strains of red salt-loving algae pearled the edges of lagoons while hairlike grass waved back and forth underwater in slow, meandering currents. We stripped off boots, pants rolled up, socks left on so we wouldn't cut our feet on frozen salt crystals. Stepping in, we broke through ice like the thinnest glass, slivers letting loose and floating downstream. For the feet, it was a stronger wake-up than a bite off the devil's nut. Plumes of salt grass grew out in the middle, and as we passed them, we spied slender, salt-crusted bones of flamingos petrified in place.

Eight flamingos cruised in overhead, landing with a flurry of pink wings along a nearby river bend. They were followed by another six that arrived from the other direction. I imagined this tableau the last concentration of bigger than microscopic life on earth as the sun burns away our atmosphere and all that remains are brine shrimps, feathery saltwater grasses, and weird pink birds.

Our steps sank through chilly water, crunching lenses of ice in the mud. We stirred up a stinky, anoxic death bin, gut-gray mud spewing from below. Invisible plumes of microbes came up with them. These are the true life out here, guaranteed survival to the bitter end. Microbes were exactly what Edward O. Wilson had warned me about when he said, "How far do you want to take it?"

No matter how rotten the circumstances, life will probably remain on this earth, or at least be buried within it. Microbes have been discovered suspended in tiny pockets of water captured within halite crystals.

Drill cores from salt flats in both the Atacama Desert and Death Valley, California, turned up viable microbes dating forty thousand years old. Given the right conditions in the lab, these ancient microbes, eaters of arsenic and magnesium, came to life again. Much older dates have been suggested from other, similar findings, even to the point of microbes surviving up to 250 million years suspended in halite. It is believed that ancient microbial ecosystems held in halite inclusions can be traced back at least a billion years. In fact, the halite found in that 4.57-billion-year-old meteorite had tiny inclusions of water inside. This water comes from the beginning of the solar system, a brine from long before earth. Though extraterrestrial microbes were not found within this meteorite-halite water, it is reasonable to think life could travel through space this way. This would support the panspermia theory, where the origins of life are thought to be held inside meteorites that strike down and release their microbes.

We trudged through putrefied, degassing bacteria. Black billows smoked the water downstream. Stepping out the other side, we stomped onto dry salt crust, our socks slick with grime. We peeled them off and wrung them out, enough microbial life right there to start up a whole other earth.

The Atacama is the loss I've lamented since the ice first began melting, since civilizations burned and mountains around me fell into rivers that carried them away. It is everything burned down. Even here at the very end, the bald earth where my own kind has ceased to exist, there was still no end. I waded on through flowing salt water around mineralized islands, cupcake-shaped protuberances of phosphates and sodium and chlorides standing above the water. Their shapes were not rough and unimaginative. Some were clumped like stromatolites, seemingly lifeless corals and staghorns. Brain-like convolutions grew right from the ground. These inventions of evaporation left me wondering if life could begin again in self-organizing formations. These islands could become home to chains of DNA coated in gossamer shells of salt, a new way of living.

Tom Neumann, one of the NASA ice researchers I'd spent time with in Greenland, told me proudly that he was part of a team who put "drill here" on the map of Antarctica where the largest ice-drilling project in history was now under way. They were six years into it, at a depth of 10,928 feet and nearing the bottom, and Neumann told me he and many other scientists and engineers on the project were concerned about breaching the underside of the ice sheet with the drill.

I didn't understand the concern, and he explained that the pressure of ice weight at that depth would close the drill hole if they didn't pump down a mixture of toxic hydrocarbons to hold it open. "If we break through, it will contaminate everything the fluid contacts," he said.

I imagined under the immense ice sheet a gush of millions upon millions of gallons of toxins, but who would care, what damage could it possibly do? Isn't the underside of ice just another obliterated world?

"There is liquid water beneath the ice sheet," Neumann reminded me.

Still, I didn't get it. All in the name of science, right?

"You of anyone should know that anywhere you find water on this planet, you're going to find life."

"Oh," I said, suddenly realizing what he had been trying to tell me.

The drill would break through into a reservoir of microbial life-forms that have possibly been in isolation for millions of years. Our first contact would be deadly, another push for extinction. Nearer the coast of Antarctica, a blood-red microbial waterfall has been found spilling from inside a glacier. Metabolizing ferric ions and sulfates (rather than oxygen), those microbes are from a lake of seawater brine that was sealed into ice about two million years ago, long enough for a unique genetic environment to evolve. Those microbes represent the potentential for a completely different future of life on earth.

Neumann's concern over contaminating such a genetic nursery is understandable. There are international laws preventing such biotic violations in Antarctica, yet he didn't seem troubled by the laws as much as he was by the idea of what damage the scientists could do.

"There's a really good science reason to be cautious," he said. "Sure, you can make a lot of interesting measurements on the sorts of physical properties of the system down there—water temperature, flow rates, heat sources, mineralogy, et cetera—but the biology questions are equally interesting and important. If we can't figure out how to study the system without having the measuring process itself fundamentally alter the system, then we're really shooting ourselves in the foot. For example, it's hard to study the properties of an eggshell if you open it with a chain saw; you'd just obliterate the thing you wanted to study."

The drill finally stopped fifty feet from the bottom of the ice, which is where the project ended. If there is something alive and tenaciously adapted down there, it has been left to itself. It may continue stewing for millions more years. Add halite crystals into the mix at some point and hold these microbes in a brine inclusion. Then, blow up the planet. You could see this life suspended inside a meteor hurtling for some other world.

The earth is a seed planting itself over and over. We are not the gardeners. We are no benevolent being leaving the house every morning with a watering can and a trowel to dig up weeds, wiping our brows midday to marvel at our handiwork. Instead, we are within the seed itself. We are part of its cells and the hardness of its coat, our place not to marvel at the futility and smallness of ourselves but to keep life moving. What we do now, from the inside, determines the vigor of that seed, how long it might live and plant itself again.

Wearing yesterday's socks (always bring at least two pair), we pulled on boots and walked out of the brine. After a good few hours traveling away from water, we stopped for our afternoon nap at a plume of saw grass growing out in the middle of a salt field. It was the only thing anywhere, a solitary burst of one green firework. This was no kipuka full of snails and fern spores, no insects I could see. But it was life, we knew the kind, something we could glide our fingers through. We crawled back into its slender, razor-like shade and waited out a bit of sun. JT's breath fell into a familiar rhythm and my heart beat through the leaves and I felt like I was made for this.

From there out it was wind-stripped boulders and coarse sand.

Our drinking water lasted to the road, perfect timing. By sunset we took our last slugs and walked along a gravel washboard coming from a lithium mine ten miles away. A Chilean man driving a white van picked us up. He was a parts supplier to the mine. We asked where he was heading. He told us San Pedro. JT and I looked at each other with a fine, earth-loving grin. Bars and bonfires, let's go.

I woke on a hard bed. The inside of my mouth felt as if something had crawled in and died. I hadn't realized until then how late our night had gone. I remembered that by the end of the night people were burning what looked like furniture, throwing pieces of a broken-up chair into the fire while JT and I ordered another round of pisco sours in frosted champagne glasses chased with more of that fine Austral Calafate beer in a half-outdoor establishment where beer bottles were then picked up for us. The fire would rise, and we'd have to scoot back another few feet. It would fall, and we would close in around it, a couple town dogs lounging at our feet.

As we wobbled back to our *hospedaje* under the yellow halos of streetlights, JT had started kicking a can and some other drunk Chilean had joined in, and I watched the two of them kick that can down the block as if they were little boys. Even out here, no, *especially* out here, at the end of the world, life unfurls.

It was late and quiet when we got back to our linoleum-floored room with its two hard beds, where we drooled and snored until the sun came up.

Both squinty eyed and slouched, we took breakfast on a patio with hollyhocks and pomegranate trees. The adobe plaza where we'd rented a room filled with sharp light. On white plates sat a perfect strawberry for each of us, and soft-crusted toast, and eggs cooked in the shell. JT smiled and said, "I like civilization."

A wind started blowing, tipping our napkins. Then somebody had to take down the shade umbrella. Then doors were closed, windows shuttered. By afternoon, a gray-clouded wind had set upon the town, twisting the imported Chinese elms and sending away dry pomegranate leaves. It turned into the kind of stiff wind the Atacama is famous for, one we were glad to have missed over the last several days.

We were in town for a resupply. Load up water, hire a ride, jump back into the desert. We talked about options, thinking a second night in town might be nice, return to the bonfire for pisco sours, maybe a good Argentine steak beforehand, something to fatten us before the next excursion. Or maybe we'd stay a few nights, or maybe we'd never get out of here, drinking and burning things until we grew donkey ears and had to stay forever.

The wind didn't smell like dust. It smelled like rain. This sweet water scent in the air seemed odd for such a famously arid place. Through blowing dust and grit I could make out the rot of pomegranates on the ground and the faint smell of hollyhocks.

I asked the owner if it might rain, and he said no. Standing in the lee of an adobe wall, we were both looking at the clouds, which I swore looked like rain clouds. He said I was mistaken. A potbellied Chilean, he had lived here all his life and said that when it does rain in this particular location, the sky has to turn silvery and the clouds come in lower with a kind of glow. He spoke as if remembering something from childhood, a legendary instant the way some people talk about tornadoes. He described an unusual metallic color to the clouds and a stillness that preceded rain, as if the world held its breath for one of the most amazing, essential features on this planet, fresh water falling from the sky.

"Not today," the man said, practically laughing in rough English. "I promise, not today."

I was disappointed, having thought we'd arrived for a historic moment.

Windstorm or not, JT and I needed to wash our clothes and get them hung and dried just in case we decided to get out of here right away. In the rear of the *hospedaje,* we rinsed laundry in a bucket, kneading and sudsing pants and shirts, wringing salt out of socks and pouring it on the ground, returning it to the desert.

As we pinned clothes to a wire, raindrops struck our faces. They were as faint as light kisses. One marked the ground, then another, dime-sized shadows dropped from the sky. Dried bougainvillea spun around the plaza as we looked up, seeing perfectly good rain clouds flying by, silvery blue and racing past in the wind.

Acknowledgments

There are many I am grateful to, above all Regan and our boys, whose lives bore every tale in this book. I brought home more dust than toys, but that, too, is love. To Dan Frank, my editor at Pantheon, I appreciate your willing candor, and your leaning across the table beaming when I first told you I wanted to write this book. Kathy Anderson, my butt-kicking agent who helped me realize this book was possible, thanks for spinning me around New York, especially for tracking fireflies with me through the city. To my early chapter readers—you know your names—I give a bow, and condolences for my panicked last-minute inquiries. In you I found community. There were also journal and magazine editors who worked on pieces of text for articles; I appreciate your scrutiny, especially that of Jennifer Sahn at *Orion,* who sat on a floor covered with pages and poured me a drink.

I cannot express enough gratitude to the scientists both named and mostly unnamed who asked the key questions and compiled data, making the vision in this book tangible. You have my deepest admiration.

Both the Ellen Meloy Fund for Desert Writers and the Rowell Award for the Art of Adventure, thank you. I hope to have honored the lives and work of Ellen, Galen, and Barbara.

Outdoor companies helped with some of these trips by donations of either cold cash or needed gear. I thank Patagonia clothing, Osprey packs, Mountain Gear, Chaco, Clif Bar, Big Agnes, and Canada Goose. (Crampons in chapter 2, the ones in the first sex scene, were generously donated by Kahtoola.)

Finally, I owe much to the town of Paonia, Colorado, near where I live, so many of its residents firsthand witnesses to the evolution of this book, their comments, quotes, and after-hours critiques were indispensable. Especially, thanks go to Revolution Brewing in Paonia—Mike grinding that grain by human power, Goold glancing from behind the bar to make sure I was still writing—without that little table on which to pile my journals and a plug for my computer, I don't know what I would have done.

Bibliography

ASTEROIDS AND VOLCANOES

Abramov, Oleg, and Stephen J. Mojzsis. "Microbial Habitability of the Hadean Earth During the Late Heavy Bombardment." *Nature* 459 (2009): 419–422.

Bottke, William F., David Vokrouhlický, and David Nesvorný. "An Asteroid Breakup 160 Myr Ago as the Probable Source of the K/T Impactor." *Nature* 449 (2007): 48–53.

Culler, Timothy S., Timothy A. Becker, Richard A. Muller, and Paul R. Rennel. "Lunar Impact History from $^{40}Ar/^{39}Ar$ Dating of Glass Spherules." *Science* 287 (2000): 1785–1788.

Darling, J., C. Storey, and C. Hawkesworth. "Impact Melt Sheet Zircons and Their Implications for the Hadean Crust." *Geology* 37 (2009): 927–930.

De Laubenfels, M. W. "Dinosaur Extinction: One More Hypothesis." *Journal of Paleontology* 30 (1956): 207–218.

Elkins-Tanton, Linda T., and Bradford H. Hager. "Giant Meteoroid Impacts Can Cause Volcanism." *Earth and Planetary Science Letters* 239 (2005): 219–232.

Gomes, R., H. F. Levison, K. Tsiganis, and A. Morbidelli. "Origin of the Cataclysmic Late Heavy Bombardment Period of the Terrestrial Planets." *Nature* 435 (2005): 466–469.

Grattona, Julio, and Carlos Alberto Perazzo. "Catastrophic Bolide Impacts on the Earth: Some Estimates." *American Journal of Physics* 74 (2006): 789–793.

Hagstrum, Jonathan T. "Antipodal Hotspots and Bipolar Catastrophes: Were Oceanic Large-Body Impacts the Cause?" *Earth and Planetary Science Letters* 236 (2005): 13–27.

Head, James W., III, Caleb I. Fassett, Seth J. Kadish, David E. Smith, Maria T. Zuber, Gregory A. Neumann, and Erwan Marzarico. "Global Distribution of Large Lunar Craters: Implications for Resurfacing and Impactor Populations." *Science* 329 (2010): 1504–1507.

Keller, G. "Impacts, Volcanism and Mass Extinction: Random Coincidence or Cause and Effect?" *Australian Journal of Earth Sciences* 52 (2005): 725–757.

Kerr, Richard A. "Did an Impact Trigger the Dinosaurs' Rise?" *Science* 296 (2002): 1215–1216.

Bibliography

Levine, Jonathan, Timothy A. Becker, Richard A. Muller, and Paul R. Renne. "$^{40}Ar/^{39}Ar$ Dating of Apollo 12 Impact Spherules." *Geophysical Research Letters* 32 (2005): L15201.

Marinova, Margarita M., Oded Aharonson, and Erik Asphaug. "Do Mega Impacts Leave Craters? Characterizing Mega Impacts and Their Relation to the Mars Hemispheric Dichotomy." Paper presented at the Seventh International Conference on Mars, Pasadena, Calif., July 9–13, 2007.

Morrow, Jared R. "Impacts and Mass Extinctions Revisited." *Palaios* 21 (2006): 313–315.

Olsen, P. E., D. V. Kent, H.-D. Sues, C. Koeberl, H. Huber, A. Montanari, E. C. Rainforth, S. J. Fowell, M. J. Szajna, and B. W. Hartline. "Ascent of Dinosaurs Linked to an Iridium Anomaly at the Triassic-Jurassic Boundary." *Science* 296 (2002): 1305–1307.

Prothero, Donald R. "Did Impacts, Volcanic Eruptions, or Climate Change Affect Mammalian Evolution?" *Palaeogeography, Palaeoclimatology, Palaeoecology* 214 (2004): 283–294.

Wells, Llyd E., John C. Armstrong, and Guillermo Gonzalez. "Reseeding of Early Earth by Impacts of Returning Ejecta During the Late Heavy Bombardment." *Icarus* 162 (2003): 38–46.

Wignall, Paul, B., Yadong Sun, David P.G. Bond, Gareth Izon, Robert J. Newton, Stéphanie Védrine, Mike Widdowson, Jason R. Ali, Xulong Lai, Haishui Jiang, Helen Cope and Simon H. Bottrell. "Volcanism, Mass Extinction, and Carbon Isotope Fluctuations in the Middle Permian of China." *Science* 29 (2009): 1179–1182.

White, Rosalind V., and Andrew D. Saunders. "Volcanism, Impact, and Mass Extinctions: Incredible or Credible Coincidences?" *Lithos* 79 (2005): 299–316.

CLIMATE CHANGE

Adams, Jonathan, Mark Maslin, and Ellen Thomas. "Sudden Climate Transitions During the Quaternary." *Progress in Physical Geography* 23 (1999): 1–36.

Alley, R. B., J. Marotzke, W. D. Nordhaus, J. T. Overpeck, D. M. Peteet, R. A. Pielke Jr., R. T. Pierrehumbert, P. B. Rhines, T. F. Stocker, L. D. Talley, and J. M. Wallace. "Abrupt Climate Change." *Science* 299 (2003): 2005–2010.

Archer, David, Michael Eby, Victor Brovkin, Andy Ridgwell, Long Cao, Uwe Mikolajewicz, Ken Caldeira, Katsumi Matsumoto, Guy Munhoven, Alvaro Montenegro, and Kathy Tokos. "Atmospheric Lifetime of Fossil Fuel Carbon Dioxide." *Annual Review of Earth and Planetary Sciences* 37 (2009): 117–134.

Broeker, Wally S., and Sidney Hemming. "Climate Swings Come into Focus." *Science* 294 (2001): 2308–2309.

Clark, Peter U., Nicklas G. Pisias, Thomas F. Stocker, and Andrew J. Weaver. "The

Role of the Thermohaline Circulation in Abrupt Climate Change." *Nature* 415 (2002): 863–869.

Eby, M., K. Zickfield, and A. J. Weaver. "Lifetime of Anthropogenic Climate Change: Millennial Time Scales of Potential CO_2 and Surface Temperature Perturbations." *Journal of Climate* 22 (2009): 2501–2511.

Grebmeier, Jacqueline M., James E. Overland, Sue E. Moore, Ed V. Farley, Eddy C. Carmack, Lee W. Cooper, Karen E. Frey, John H. Helle, Fiona A. McLaughlin, and S. Lyn McNutt. "A Major Ecosystem Shift in the Northern Bering Sea." *Science* 311 (2006): 1461–1464.

Grimm, Nancy B., Stanley H. Faeth, Nancy E. Golubiewski, Charles L. Redman, Jianguo Wu, Xuemei Bai, and John M. Briggs. "Global Change and the Ecology of Cities." *Science* 319 (2008): 756–760.

Gupta, Anil K., David M. Anderson, and Jonathan T. Overpeck. "Abrupt Changes in the Asian Southwest Monsoon During the Holocene and Their Links to the North Atlantic Ocean." *Nature* 421 (2003): 354–357.

Hansen, James, Larissa Nazarenko, Reto Ruedy, Makiko Sato, Josh Willis, Anthony Del Genio, Dorothy Kick, Andrew Lacis, Ken Lo, Surabi Menon, Tica Novakov, Judith Perlwitz, Gary Russel, Gavin A. Schmidt, and Nicholas Tausnev. "Earth's Energy Imbalance: Confirmation and Implications." *Science* 308 (2005): 1431–1435.

Hu, Y., and Q. Fu. "Observed Poleward Expansion of the Hadley Circulation Since 1979." *Atmospheric Chemistry and Physics* 7 (2007): 21–24.

Johanson, Celeste M., and Qiang Fu. "Hadley Cell Widening: Model Simulations Versus Observations." *Journal of Climate* 22 (2009): 2713–2725.

Lu, Jian, Gabriel A. Vecchi, and Thomas Reichler. "Expansion of the Hadley Cell Under Global Warming." *Geophysical Research Letters* 34 (2007): 2–6.

Overland, James E., and Phyllis J. Stabeno. "Is the Climate of the Bering Sea Warming and Affecting the Ecosystem?" *EOS* 85 (2004): 309–316.

Post, John D. *The Last Great Subsistence Crisis in the Western World*. Baltimore: Johns Hopkins University Press, 1977.

Rial, José A. "Abrupt Climate Change: Chaos and Order at Orbital and Millennial Scales." *Global and Planetary Change* 41 (2004): 95–109.

Rial, José A., Roger A. Pielke, Martin Beniston, Martin Claussen, Josep Canadell, Peter Cox, Hermann Held, Nathalie de Noblet-Ducoudré, Ronald Prinn, and James F. Reynolds. "Nonlinearities, Feedbacks, and Critical Thresholds Within the Earth's Climate System." *Climate Change* 65 (2004): 11–38.

Root, Terry L., Jeff T. Price, Kimberly R. Hall, Stephen H. Schnelder, Cynthia Rosenzweig, and J. Alan Pounds. "Fingerprints of Global Warming on Wild Animals and Plants." *Nature* 421 (2003): 57–60.

Ruddiman, William F. "The Anthropogenic Greenhouse Era Began Thousands of Years Ago." *Climatic Change* 61 (2003): 261–293.

Scheffer, Marten, Jordi Bascompte, William A. Brock, Victor Brovkin, Stephen R. Carpenter, Vasilis Dakos, Hermann Held, Egbert H. van Nes, Max Rietkerk, and George Sugihara. "Early-Warning Signals for Critical Transitions." *Nature* 461 (2009): 53–59.

Scheffer, Marten, Steve Carpenter, Jonathan A. Foley, Carl Folkes, and Brian Walker. "Catastrophic Shifts in Ecosystems." *Nature* 413 (2001): 591–596.

Shaffer, Gary. "Long Time Management of Fossil Fuel Resources to Limit Global Warming and Avoid Ice Age Onsets." *Geophysical Research Letters* 36 (2009): L03704.

Shaw, Justine M. "Climate Change and Deforestation: Implications for the Maya Collapse." *Ancient Mesoamerica* 14 (2003): 157–167.

Sidel, Dian J., Qiang Fu, William J. Randel, and Thomas J. Reichler. "Widening of the Tropical Belt in a Changing Climate." *Nature Geoscience* 1 (2008): 21–24.

Wesenbeeck, Bregje K. van, Johan van de Koppel, Peter M. J. Herman, Mark D. Bertness, Daphne van der Wal, Jan P. Bakker, and Tjeerd J. Bouma. "Potential for Sudden Shifts in Transient Systems: Distinguishing Between Local and Landscape-Scale Processes." *Ecosystems* 11 (2008): 1133–1141.

Zachos, James, Mark Pagani, Lisa Sloan, Ellen Thomas, and Katharina Billups. "Trends, Rhythms, and Aberrations in Global Climate 65 Ma to Present." *Science* 292 (2001): 686–693.

DESERTS

Ackland, Graeme J., Michael A. Clark, and Timothy M. Lenton. "Catastrophic Desert Formation in Daisyworld." *Journal of Theoretical Biology* 223 (2003): 39–44.

Fawcett, Peter J., Josef P. Werne, R. Scott Anderson, Jeffrey M. Heikoop, Erik T. Brown, Melissa A. Berke, Susan J. Smith, Fraser Goff, Linda Donohoo-Hurley, Luz M. Cisneros-Dozal, Stefan Schouten, Jaap S. Sinninghe Damsté, Yongsong Huang, Jaime Toney, Julianna Fessenden, Giday WoldeGabriel, Viorel Atudorei, John W. Geissman, and Craig D. Allen. "Extended Megadroughts in the Southwestern United States During Pleistocene Interglacials." *Nature* 470 (2011): 518–521.

Geist, Helmut J., and Eric F. Lambin. "Dynamic Causal Patterns of Desertification." *BioScience* 54 (2004): 817–829.

Hansen, Zeynep K., and Gary D. Libecap. "Small Farms, Externalities, and the Dust Bowl of the 1930s." *Journal of Political Economy* 112 (2004): 665–694.

Jickells, T. D., Z. S. An, K. K. Andersen, A. R. Baker, G. Bergametti, N. Brooks, J. J. Cao, P. W. Boyd, N. Mahowald, J. M. Prospero, A. J. Ridgwell, I. Tegen, and R. Torres. "Global Iron Connections Between Desert Dust, Ocean Biogeochemistry, and Climate." *Science* 308 (2005): 67–71.

Miao, Xiaodong, Joseph A. Mason, James B. Swinehart, David B. Loope, Paul R.

Hanson, Ronald J. Goble, and Xiaodong Liu. "A 10,000 Year Record of Dune Activity, Dust Storms, and Severe Drought in the Central Great Plains." *Geology* 35 (2007): 119–122.

Mulitza, Stefan, David Heslop, Daniela Pittauerova, Helmut W. Fischer, Inka Meyer, Jan-Berend Stuut, Matthias Zabel, Gesine Mollenhauer, James A. Collins, Henning Kuhnert, and Michael Schulz. "Increase in African Dust Flux at the Onset of Commercial Agriculture in the Sahel Region." *Nature* 466 (2010): 226–228.

Prospero, Joseph M., and Peter J. Lamb. "African Droughts and Dust Transport to the Caribbean: Climate Change Implications." *Science* 302 (2003): 1024–1027.

Schubert, Siegfried D., Max J. Suarez, Philip J. Pegion, Randal D. Koster, and Julio T. Bacmeister. "On the Cause of the 1930s Dust Bowl." *Science* 303 (2004): 1855–1859.

Sridhar, Venkataramana, David B. Loope, James B. Swinehart, Joseph A. Mason, Robert K. Oglesby, and Clinton M. Rowe. "Large Wind Shift on the Great Plains During the Medieval Warm Period." *Science* 21 (2006): 345–347.

Wolfe, Stephen A., and Christopher H. Hugenholtz. "Barchan Dunes Stabilized Under Recent Climate Warming on the Northern Great Plains." *Geology* 37 (2009): 1039–1042.

EXTINCTIONS (CURRENT)

Barnosky, Anthony D., Nicholas Matzke, Susumu Tomiya, Guinevere O. U. Wogan, Brian Swartz, Tiago B. Quental, Charles Marshall, Jenny L. McGuire, Emily L. Lindsey, Kaitlin C. Maguire, Ben Mersey, and Elizabeth A. Ferrer. "Has the Earth's Sixth Mass Extinction Already Arrived?" *Nature* 471 (2011): 51–57.

Ceballos, Gerardo, and Paul R. Ehrlich. "Mammal Population Losses and the Extinction Crisis." *Science* 296 (2002): 904–907.

Drake, John M., and Blaine D. Griffen. "Early Warning Signals of Extinction in Deteriorating Environments." *Nature* 467 (2010): 456–459.

Gastaldo, Robert A., Johann Neveling, C. Kittinger Clark, and Sophia S. Newbury. "The Terrestrial Permian-Triassic Boundary Event Bed Is a Nonevent." *Geology* 27 (2009): 199–202. (In response: Peter D. Ward, Gregory J. Retallack, and Roger M. H. Smith, "The Terrestrial Permian-Triassic Boundary Event Bed Is a Nonevent: Comment," *Geology* 40 [2012]: e256.)

Gathorne-Hardy, F. J., and W. E. H. Harcourt-Smith. "The Super-Eruption of Toba, Did It Cause a Human Bottleneck?" *Journal of Human Evolution* 45 (2003): 227–230. (In response: Stanley H. Ambrose, "Did the Super-Eruption of Toba Cause a Human Population Bottleneck? Reply to Gathorne-Hardy and Harcourt-Smith," *Journal of Human Evolution* 45 [2003]: 231–237.)

He, Fangliang, and Stephen P. Hubbell. "Species-Area Relationships Always Overestimate Extinction Rates from Habitat Loss." *Nature* 473 (2011): 368–371.

McCallum, Malcolm L. "Amphibian Decline or Extinction? Current Declines Dwarf Background Extinction Rate." *Journal of Herpetology* 41 (2007): 483–491.

McKinnon, Jeffrey S., and Eric B. Taylor. "Biodiversity: Species Choked and Blended." *Nature* 482 (2012): 313–314.

Sax, Dov F., and Steven D. Gaines. "Species Diversity: From Global Decreases to Local Increases." *Trends in Ecology and Evolution* 18 (2003): 561–566.

Stork, Nigel E. "Re-assessing Current Extinction Rates." *Biodiversity & Conservation* 19 (2009): 357–371.

Sutherland, William J. "Parallel Extinction Risk and Global Distribution of Languages and Species." *Nature* 423 (2003): 276–279.

Tilman, David, Clarence L. Lehman, and Chengjun Yin. "Habitat Destruction, Dispersal, and Deterministic Extinction in Competitive Communities." *American Naturalist* 149 (1997): 407–435.

Wake, David B., and Vance T. Vredenburg. "Are We in the Midst of the Sixth Mass Extinction? A View from the World of Amphibians." *Proceedings of the National Academy of Sciences of the United States of America* 105 (2008): 11466–11473.

White, Rosalind V., and Andrew D. Saunders. "Volcanism, Impact, and Mass Extinctions: Incredible or Credible Coincidences?" *Lithos* 79 (2005): 299–316.

Wiens, Delbert, and Michèle R. Slaton. "The Mechanism of Background Extinction." *Biological Journal of the Linnean Society* 105 (2012): 255–316.

EXTINCTIONS (ANCIENT)

Alvarez, Luis W., Walter Alvarez, Frank Asaro, and Helen V. Michel. "Extraterrestrial Cause for the Cretaceous-Tertiary Extinction." *Science* 208 (1980): 1095–1108.

Benton, Michael J., and Richard J. Twitchett. "How to Kill (Almost) All Life: The End-Permian Extinction Event." *Trends in Ecology and Evolution* 18 (2003): 358–365.

Ceballos, Gerardo, and Paul R. Ehrlich. "Mammal Population Losses and the Extinction Crisis." *Science* 296 (2002): 904–907.

Fassett, James E. "The Documentation of In-Place Dinosaur Fossils in the Paleocene Ojo Alamo Sandstone and Animas Formation in the San Juan Basin of New Mexico and Colorado Mandates a Paradigm Shift: Dinosaurs Can No Longer Be Thought of as Absolute Index Fossils for End-Cretaceous Strata in the Western Interior of North America." *New Mexico Geology* 29 (2007): 56.

Fassett, James E., Larry M. Heaman, and Antonio Simonetti. "Direct U-Pb Dating of Cretaceous and Paleocene Dinosaur Bones, San Juan Basin, New Mexico." *Geology* 39 (2011): 159–162.

Grasby, Stephen E., Hamed Sanei, and Benoit Beauchamp. "Catastrophic Dispersion of Coal Fly Ash into Oceans During the Latest Permian Extinction." *Nature Geoscience* 4 (2011): 104–107.

Huey, Raymond B., and Peter D. Ward. "Hypoxia, Global Warming, and Terrestrial Late Permian Extinctions." *Science* 308 (2005): 398–401.

Kiehl, Jeffrey T. and Christine A. Shields. "Climate Simulation of the Latest Permian: Implications for Mass Extinction." *Geology* 33 (2005): 757–760.

Ogden, Darcy E., and Norman H. Sleep. "Explosive Eruption of Coal and Basalt and the End-Permian Mass Extinction." *Proceedings of the National Academy of Sciences of the United States of America* 109 (2012): 59–62.

Payne, Jonathan L., and Matthew E. Clapham. "End-Permian Mass Extinction in the Oceans: An Ancient Analog for the Twenty-First Century?" *Annual Review of Earth and Planetary Sciences* 40 (2012). doi:10.1146.

Payne, Jonathan L., Daniel J. Lehrmann, Jiayong Wei, Michael J. Orchard, Daniel P. Schrag, and Andrew H. Knoll. "Large Perturbations of the Carbon Cycle During Recovery from the End-Permian Extinction." *Science* 305 (2004): 506–509.

Peters, Shanan E. "Environmental Determinants of Extinction Selectivity in the Fossil Record." *Nature* 454 (2008): 626–629.

White, Rosalind V. "Earth's Biggest 'Whodunnit': Unravelling the Clues in the Case of the End-Permian Mass Extinction." *Philosophical Transactions of the Royal Society* 360 (2002): 2963–2985.

ICE

Barnett, T. P., J. C. Adam, and D. P. Lettenmaier. "Potential Impacts of a Warming Climate on Water Availability in Snow-Dominated Regions." *Nature* 438 (2005): 303–309.

Benn, Douglas I. "Glaciers." *Progress in Physical Geography* 30 (2006): 432–444.

Caputo, Mário V., and John C. Crowell. "Migration of Glacial Centers Across Gondwana During Paleozoic Era." *Geologic Society of America Bulletin* 96 (1985): 1020–1036.

Carey, Mark. "Living and Dying with Glaciers: People's Historical Vulnerability to Avalanches and Outburst Floods in Peru." *Global and Planetary Change* 47 (2004): 122–134.

Chen, J. L., C. R. Wilson, D. Blankenship, and B. D. Tapley. "Accelerated Antarctic Ice Loss from Satellite Gravity Measurements." *Nature Geoscience* (2009). doi:10.1038/NGEO694.

Chen, J. L., C. R. Wilson, B. D. Tapley, D. D. Blankenship, and E. R. Ivins. "Patagonia Ice Field Melting Observed by Gravity Recovery and Climate Experiment (GRACE)." *Geophysical Research Letters* 34 (2007): L22501.

Dussaillant, Alejandro, Gerardo Benito, Wouter Buytaert, Paul Carling, Claudio Meier, and Fabián Espinoza. "Repeated Glacial-Lake Outburst Floods in Patagonia: An Increasing Hazard?" *Natural Hazards* 54 (2009): 469–481.

Hoffman, Paul F., and Daniel P. Schrag. "The Snowball Earth Hypothesis: Testing the Limits of Global Change." *Terra Nova* 14 (2002): 129–155.

Klemann, Volker, Erik R. Ivins, Zdenek Martinec, and Detlef Wolf. "Models of Active Glacial Isostasy Roofing Warm Subduction: Case of the South Patagonian Ice Field." *Journal of Geophysical Research* 112 (2006): B09405.

Oerlemans, J. "Extracting a Climate Signal from 169 Glacier Records." *Science* 308 (2005): 675–677.

Overpeck, Jonathan. "Arctic System on Trajectory to New, Seasonally Ice-Free State." *Eos* 86 (2005): 309–313.

Paul, Frank, Andreas Kääb, Max Maisch, Tobias Kellenberger, and Wilfried Haeberli. "Rapid Disintegration of Alpine Glaciers Observed with Satellite Data." *Geophysical Research Letters* 31 (2004): L21402.

Pollard, David, and Robert M. DeConto. "Modelling West Antarctic Ice Sheet Growth and Collapse Through the Past Five Million Years." *Nature* 458 (2009): 329–332.

Rack, Wolfgang, and Helmut Rott. "Pattern of Retreat and Disintegration of the Larsen B Ice Shelf, Antarctic Peninsula." *Annals of Glaciology* 39 (2004): 505–510.

Regan, Helen M., Richard Lupia, Andrew N. Drinnan, and Mark A. Burgman. "The Currency and Tempo of Extinction." *American Naturalist* 157 (2001): 1–10.

Ridley, J. K., P. Huybrechts, J. M. Gregory, and J. A. Lowe. "Elimination of the Greenland Ice Sheet in a High CO_2 Climate." *Journal of Climate* 18 (2005): 3409–3427.

Rignot, E., G. Casassa, P. Gogineni, W. Krabill, A. Rivera, and R. Thomas. "Accelerated Ice Discharge from the Antarctic Peninsula Following the Collapse of Larsen B Ice Shelf." *Geophysical Research Letters* 31 (2004): L18401.

Rignot, Eric, Michele Koppes, and Isabella Velicogna. "Rapid Submarine Melting of the Calving Faces of West Greenland Glaciers." *Nature Geoscience* 3 (2010): 187–191.

Rivera, Andrés, Toby Benham, Gino Casassa, Jonathan Bamber, and Julian A. Dowdeswell. "Ice Elevation and Areal Changes of Glaciers from the Northern Patagonia Icefield, Chile." *Global and Planetary Change* 59 (2007): 126–137.

Rupper, Summer, and Gerard Roe. "Glacier Changes and Regional Climate: A Mass and Energy Balance Approach." *Journal of Climate* 21 (2008): 5384–5401.

Scambos, Ted, Christina Hulbe, and Mark Fahnestock. "Climate-Induced Ice Shelf Disintegration in the Antarctic Peninsula." *Paleobiology and Paleoenvironments of Eocene Rocks and Antarctic Research Series* 76 (2003): 335–347.

Schrag, Daniel P., Robert A. Berner, Paul F. Hoffman, and Galen P. Halverson. "On the Initiation of a Snowball Earth." *Geochemistry Geophysics Geosystems* 3 (2002). doi:10.1029/2001GC000219.

Shepherd, Andrew, Duncan Wingham, Tony Payne, and Pedro Skvarca. "Larsen Ice Shelf Has Progressively Thinned." *Science* 302 (2003): 856–859.

Tarasov, Lev, and W. R. Peltier. "Arctic Freshwater Forcing of the Younger Dryas Cold Reversal." *Nature* 435 (2005): 662–665.

———. "Greenland Glacial History, Borehole Constraints, and Eemian Extent." *Journal of Geophysical Research* 108 (2003): 2143–2163.

Teller, James T., David W. Leverington, and Jason D. Mann. "Freshwater Outbursts to the Oceans from Glacial Lake Agassiz and Their Role in Climate Change During the Last Deglaciation." *Quaternary Science Reviews* 21 (2002): 879–887.

Turner, J. S. "The Melting of Ice in the Arctic Ocean: The Influence of Double-Diffusive Transport of Heat from Below." *Journal of Physical Oceanography* 40 (2010): 249–256.

Van den Broeke, Michiel. "Strong Surface Melting Preceded Collapse of Antarctic Peninsula Ice Shelf." *Geophysical Research Letters* 32 (2005): L12815.

Xu, Jianchu, R. Edward Grumbine, Arun Shrestha, Mats Eriksson, Xuefei Yang, Yun Wang, and Andreas Wilkes. "The Melting Himalayas: Cascading Effects of Climate Change on Water, Biodiversity, and Livelihoods." *Conservation Biology* 23 (2009): 520–530.

ICE AGES

Archer, David, and Andrey Ganopolski. "A Movable Trigger: Fossil Fuel CO_2 and the Onset of the Next Glaciation." *Geochemistry Geophysics Geosystems* 6 (2005). doi:10.1029/2004GC000891.

Barber, D. C., A. Dyke, C. Hillaire-Marcel, A. E. Jennings, J. T. Andrews, M. W. Kerwin, G. Bilodeau, R. McNeely, J. Southon, M. D. Morehead, and J. M. Gagnon. "Forcing of the Cold Event of 8,200 Years Ago by Catastrophic Drainage of Laurentide Lakes." *Nature* 400 (1999): 344–348.

Bjornstad, Bruce N., Karl R. Fecht, and Christopher J. Pluhar. "Long History of Pre-Wisconsin, Ice Age Cataclysmic Floods: Evidence from Southeastern Washington State." *Journal of Geology* 109 (2001): 695–713.

Clarke, Garry K. C., Andrew B. G. Bush, and John W. M. Bush. "Freshwater Discharge, Sediment Transport, and Modeled Climate Impacts of the Final Drainage of Glacial Lake Agassiz." *Journal of Climate* 22 (2009): 2161–2180.

Coope, G. R. "Several Million Years of Stability Among Insect Species Because of, or in Spite of, Ice Age Climatic Instability?" *Philosophical Transactions of the Royal Society* 359 (2004): 209–214.

Dickey, Jean O., Steven L. Marcus, Olivier de Viron, and Ichiro Fukumori. "Recent Earth Oblateness Variations: Unraveling Climate and Postglacial Rebound Effects." *Science* 298 (2002): 1975–1977.

Greve, Ralf, Karl-Heinz Wyrwoll, and Anton Eisenhauer. "Deglaciation of the Northern Hemisphere at the Onset of the Eemian and Holocene." *Annals of Glaciology* 28 (1999): 1–8.

Hewitt, Godfrey. "Genetic Consequences of Climatic Oscillations in the Quaternary." *Philosophical Transactions of the Royal Society* 359 (2004): 183–195.

———. "The Genetic Legacy of the Quaternary Ice Ages." *Nature* 405 (2000): 907–913.

———. "Some Genetic Consequences of Ice Ages, and Their Role in Divergence and Speciation." *Biological Journal of the Linnean Society* 58 (1996): 247–276.

Lawrence, David M., Andrew G. Slater, Robert A. Tomas, Marika M. Holland, and Clara Deser. "Accelerated Arctic Land Warming and Permafrost Degradation During Rapid Sea Ice Loss." *Geophysical Research Letters* 35 (2008): L11506.

Lister, Adrian M. "The Impact of Quaternary Ice Ages on Mammalian Evolution." *Philosophical Transactions of the Royal Society* 359 (2004): 221–241.

Mangerud, Jan, Martin Jakobsson, Helena Alexanderson, Valery Astakhov, Garry K. C. Clarke, Mona Henriksen, Christian Hjort, Gerhard Krinner, Juha-Pekka Lunkka, Per Möller, Andrew Murray, Olga Nikolskaya, Matti Saarnisto, and John Inge Svendsen. "Ice-Dammed Lakes and Rerouting of the Drainage of Northern Eurasia During the Last Glaciation." *Quaternary Science Reviews* 23 (2004): 1313–1332.

Margari, V., L. C. Skinner, P. C. Tzedakis, A. Ganopolski, M. Vautravers, and N. J. Shackleton. "The Nature of Millennial-Scale Climate Variability During the Past Two Glacial Periods." *Nature Geoscience* 3 (2010): 127–131.

Marshall, Charles R., and David K. Jacobs. "Flourishing After the End-Permian Mass Extinction." *Science* 325 (2009): 1079–1080.

Normark, William R., and Jane A. Reid. "Extensive Deposits on the Pacific Plate from Late Pleistocene North American Glacial Lake Outbursts." *Journal of Geology* 111 (2003): 617–637.

Raymo, Maureen E., and Peter Huybers. "Unlocking the Mysteries of the Ice Ages." *Nature* 451 (2008): 284–285.

Severinghaus, Jeffrey P., and Edward J. Brook. "Abrupt Climate Change at the End of the Last Glacial Period Inferred from Trapped Air in Polar Ice." *Science* 286 (1999): 930–934.

Bibliography

MICROBES

Lowenstein, Tim K., Brian A. Schubert, and Michael N. Timofeeff. "Microbial Communities in Fluid Inclusions and Long-Term Survival in Halite." *GSA Today* 21 (2011): 4–9.

Navarro-González, Rafael, Fred A. Rainey, Paola Molina, Danielle R. Bagaley, Becky J. Hollen, José de la Rosa, Alanna M. Small, Richard C. Quinn, Frank J. Grunthaner, Luis Cáceres, Benito Gomez-Silva, and Christopher P. McKay. "Mars-Like Soils in the Atacama Desert, Chile, and the Dry Limit of Microbial Life." *Science* 302 (2003): 1018–1021.

Sankaranarayanan, Krithivasan, Michael N. Timofeeff, Rita Spathis, Tim K. Lowenstein, J. Koji Lum. "Ancient Microbes from Halite Fluid Inclusions: Optimized Surface Sterilization and DNA Extraction." *PLoS ONE* 6 (2011): e20683.

Schubert, Brian A., Tim K. Lowenstein, and Michael N. Timofeeff. "Microscopic Identification of Prokaryotes in Modern and Ancient Halite, Saline Valley and Death Valley, California." *Astrobiology* 9 (2009): 467–482.

Schubert, Brian A., Tim K. Lowenstein, Michael N. Timofeeff, and Matthew A. Parker. "How Do Prokaryotes Survive in Fluid Inclusions in Halite for 30 k.y.?" *Geology* 37 (2009): 1059–1062.

Wierzchos, Jacek, Carmen Ascaso, and Christopher McKay. "Endolithic Cyanobacteria in Halite Rocks from the Hyperarid Core of the Atacama Desert." *Astrobiology* 6 (2006): 415–422.

OCEANS

Blanchon, Paul, and John Shaw. "Reef Drowning During the Last Deglaciation: Evidence for Catastrophic Sea Level Rise and Ice-Sheet Collapse." *Geology* 23 (1995): 4–8.

Blum, Michael D., and Harry H. Roberts. "Drowning of the Mississippi Delta Due to Insufficient Sediment Supply and Global Sea-Level Rise." *Nature Geoscience* 2 (2009): 488–491.

Church, John A., and Neil J. White. "A 20th Century Acceleration in Global Sea-Level Rise." *Geophysical Research Letters* 33 (2006): L01602.

Cuffey, Kurt M., and Shawn J. Marshall. "Substantial Contribution to Sea-Level Rise During the Last Interglacial from the Greenland Ice Sheet." *Nature* 404 (2000): 591–594.

Ericson, Jason P., Charles J. Vörösmarty, S. Lawrence Dingman, Larry G. Ward, and Michel Meybeck. "Effective Sea-Level Rise and Deltas: Causes of Change and Human Dimension Implications." *Global and Planetary Change* 50 (2006): 63–82.

Faught, Michael K. "The Underwater Archaeology of Paleolandscapes, Apalachee Bay, Florida." *American Antiquity* 69 (2004): 275–289.

Fedje, Daryl W., and Heiner Josenhans. "Drowned Forests and Archaeology on the Continental Shelf of British Columbia, Canada." *Geology* 28 (2000): 99–102.

Hallam, A., P. B. Wignall. "Mass Extinctions and Sea-Level Changes." *Earth-Science Reviews* 48 (1999): 217–250.

Hearty, Paul J., John T. Hollin, A. Conrad Neumann, Michael J. O'Leary, and Malcolm McCulloch. "Global Sea-Level Fluctuations During the Last Interglaciation (MIS 5e)." *Quaternary Science Reviews* 26 (2007): 2090–2112.

Jevrejeva, S., J. C. Moore, A. Grinsted, and P. L. Woodworth. "Recent Global Sea Level Acceleration Started over 200 Years Ago?" *Geophysical Research Letters* 35 (2008): L08715.

Keigwin, Lloyd D., Jeffrey P. Donnelly, Mea S. Cook, Neal W. Driscoll, and Julie Brigham-Grette. "Rapid Sea-Level Rise and Holocene Climate in the Chukchi Sea." *Geology* 34 (2006): 861–864.

Lyman, J. M., S. A. Good, V. V. Gouretski, M. Ishii, G. C. Johnson, M. D. Palmer, D. M. Smith, and J. K. Willis. "Robust Warming of the Global Upper Ocean." *Nature* 465 (2010): 334–337.

Overpeck, Jonathan T., Bette L. Otto-Bliesner, Gifford H. Miller, Daniel R. Muhs, Richard B. Alley, and Jeffrey T. Kiehl. "Paleoclimatic Evidence for Future Ice-Sheet Instability and Rapid Sea-Level Rise." *Science* 311 (2006): 1747–1750.

Purkey, Sarah G., and Gregory C. Johnson. "Warming of Global Abyssal and Deep Southern Ocean Waters Between the 1990s and 2000s: Contributions to Global Heat and Sea Level Rise Budgets." *Journal of Climate* 23 (2010): 6336–6351.

Rahmstorf, Stefan. "A Semi-Empirical Approach to Projecting Future Sea-Level Rise." *Science* 315 (2007): 368–370.

Stanley, Jean-Daniel, Franck Goddio, Thomas F. Jorstad, and Gerard Schnepp. "Submergence of Ancient Greek Cities off Egypt's Nile Delta—a Cautionary Tale." *GSA Today* 14 (2004): 4–10.

Wunsch, Carl. "Towards Understanding the Paleocean." *Quaternary Science Reviews* 29 (2010): 1960–1967.

TECTONICS

Bartoli, G., M. Sarnthein, M. Weinelt, H. Erlenkeuser, D. Garbe-Schönberg, and D. W. Lea. "Final Closure of Panama and the Onset of Northern Hemisphere Glaciation." *Earth and Planetary Science Letters* 237 (2005): 33–44.

Bookhagen, Bodo, Rasmus C. Thiede, and Manfred R. Strecker. "Late Quaternary Intensified Monsoon Phases Control Landscape Evolution in the Northwest Himalaya." *Geology* 33 (2005): 149–152.

Bibliography

Breuer, D., and T. Spohn. "Early Plate Tectonics Versus Single-Plate Tectonics on Mars: Evidence from Magnetic Field History and Crust Evolution." *Journal of Geophysical Research* 108 (2003): 11969–11979.

Grujic, Djordje, Isabelle Cout, Bodo Bookhagen, Stephane Bonnet, Ann Blythe, and Chris Duncan. "Climatic Forcing of Erosion, Landscape, and Tectonics in the Bhutan Himalayas." *Geology* 34 (2006): 801–804.

Haug, Gerald H., Ralf Tiedemann, Rainer Zahn, and A. Christina Ravelo. "Role of Panama Uplift on Oceanic Freshwater Balance." *Geology* 29 (2001): 207–210.

Iaffaldano, Giampiero, Laurent Husson, and Hans-Peter Bunge. "Monsoon Speeds Up Indian Plate Motion." *Earth and Planetary Science Letters* 304 (2011): 503–510.

Lear, Caroline H., Yair Rosenthal, and James D. Wright. "The Closing of a Seaway: Ocean Water Masses and Global Climate Change." *Earth and Planetary Science Letters* 210 (2003): 425–436.

Lindsay, John F., and Martin D. Brasier. "Did Global Tectonics Drive Early Biosphere Evolution? Carbon Isotope Record from 2.6 to 1.9 Ga Carbonates of Western Australian Basins." *Precambrian Research* 114 (2002): 1–34.

Liu, X., and Z. Yin. "Sensitivity of East Asian Monsoon Climate to the Uplift of the Tibetan Plateau." *Palaeogeography, Palaeoclimatology, Palaeoecology* 183 (2002): 223–245.

Molnar, Peter, and Philip England. "Late Cenozoic Uplift of Mountain Ranges and Global Climate Change: Chicken or Egg?" *Nature* 346 (1990): 29–34.

Molnar, Peter, Philip England, and Joseph Martinod. "Mantle Dynamics, Uplift of the Tibetan Plateau, and the Indian Monsoon." *Reviews of Geophysics* 31 (1993): 357–396.

Pullen, Alex, Paul Kapp, Andrew T. McCallister, Hong Chang, George E. Gehrels, Carmala N. Garzione, Richard V. Heermance, and Lin Ding. "Qaidam Basin and Northern Tibetan Plateau as Dust Sources for the Chinese Loess Plateau and Paleoclimatic Implications." *Geology* 39 (2011): 1031–1034.

Whipple, Kelin X. "The Influence of Climate on the Tectonic Evolution of Mountain Belts." *Nature Geoscience* 2 (2009): 97–104.

Zhisheng, An, John E. Kutzbach, Warren L. Prell, and Stephen C. Porter. "Evolution of Asian Monsoons and Phased Uplift of the Himalaya-Tibetan Plateau Since Late Miocene Times." *Nature* 411 (2001): 62–66.

Craig Childs is a commentator for NPR's *Morning Edition,* and his work has appeared in *The New York Times,* the *Los Angeles Times, Men's Journal, Outside, The Sun,* and *Orion.* Awards he has won include the Ellen Meloy Desert Writers Award, the Rowell Award for the Art of Adventure, the Sigurd F. Olson Nature Writing Award, and, for his body of work, the 2003 Spirit of the West Award.

A NOTE ON THE TYPE

This book was set in Perrywood, a typeface designed by Johannes Birkenbach for the Monotype Foundry in 1993. Birkenbach, a German designer, set out to create a face that could combine the character of traditional metal types with the greater flexibility of digital types. Based loosely on typefaces such as Bembo and Plantin, Perrywood contains some of the distinctive old-style letter shapes, but with more regular proportions and weights to improve legibility in a wide range of sizes.

Composed by North Market Street Graphics,
Lancaster, Pennsylvania

Printed and bound by Berryville Graphics,
Berryville, Virginia

Book design by Robert C. Olsson